D1223600

Rheometry:
Industrial Applications

MATERIALS SCIENCE RESEARCH STUDIES SERIES

Series Editor: **Professor C. R. Tottle,** M.Met., M.Sc., F.I.M., F.Inst.P.

1. Rheometry: Industrial Applications

Edited by **Professor Kenneth Walters**

Rheometry: Industrial Applications

Edited by

Professor Kenneth Walters

Department of Applied Mathematics,
University College of Wales,
Aberystwyth, Dyfed, Wales

RESEARCH STUDIES PRESS
A DIVISION OF JOHN WILEY & SONS LTD
Chichester · New York · Brisbane · Toronto

RESEARCH STUDIES PRESS

Editorial Office:
8 Willian Way, Letchworth, Herts SG6 2HG, England

British Library Cataloguing in Publication Data:

Rheometry—(Materials science research studies series; vol. I).
　　1. Engineering instruments
　　2. Rheometers
　　I. Walters, Kenneth　　II. Series
　　620.1'064　　　TA165　　　80-40956

　　ISBN 0 471 27878 5

Printed in the United States of America

PREFACE

The 1975 BSI Glossary of Rheological Terms [1] defines a Viscometer as "an instrument for the measurement of viscosity". It is now common knowledge that, although viscosity remains the most important fluid property, there are others which demand attention both from fundamental and practical standpoints. Accordingly, in considering rheologically-complex fluids, it is more convenient to think in terms of a Rheometer (defined as "an instrument for measuring rheological properties") which clearly includes the Viscometer as a special case. The associated science is called Rheometry.

It was my original intention to write a monograph on "Rheometry : Industrial Applications" which was to be a companion volume to my "Rheometry" [2] published by Chapman and Hall in 1975. The earlier book essentially answered the "How?" question in the sense that it discussed in some detail the mechanics of making rheometrical measurements. The second book was to address itself to the "Why?" question within an industrial framework, asking such questions as "How much Rheometry is carried out in the various industries?" "What use is made of the data when these become available?" "What level of sophistication does Industry require?" "Is there a strong correlation between research laboratory rheometry and shop-floor activity?"

Some honest reflection on the magnitude of the task of becoming a short-term expert in a wide range of industries soon indicated that the project was neither feasible nor fair on the general scientific public. Accordingly, plans were revised and broadened to include a number of experts in industrial rheology and rheometry who were known to me personally and who were willing to take part in the proposed project. I undertook to ensure that, as far as possible, the end product would be a unified whole with a common notation.

It was agreed to restrict attention to rheologically complex materials that are basically fluid-like, although this was seen as including systems with a yield

v

stress.

The original intention was to provide one numbered list of references. However, it transpired that the amount of overlap in this regard in the lists provided by the individual contributors was not significant. It was accordingly decided to list the references chapter by chapter.

Much of my writing and editorial commitments were undertaken during a visit to the University of Wisconsin where I held an Honorary Fellowship in the Mathematics Research Center during the Summer of 1979. Very helpful discussions with Professor R.B. Bird and Professor A.S. Lodge of the Rheology Research Center of that University are gratefully acknowledged.

Finally, I would like to express my appreciation to Mr. R.E. Evans for his help in assembling the many figures and photographs in the book and to Mrs. D.J. Vincent for her patience and expertise in typing the camera-ready copy.

Ken Walters

December 1979

LIST OF CONTRIBUTORS

Dr. H.A. Barnes,
  Unilever Research Laboratory,
    Port Sunlight, Cheshire, U.K.

Dr. K.M. Beazley,
  E.E.C. International,
    St. Austell, Cornwall, U.K.

Mr. J.F. Hutton,
  Shell Research Limited,
    Thornton Research Centre, Chester, U.K.

Professor J. Mewis,
  Institute of Chemical Engineering,
    Catholic University of Leuven, Belgium.

Dr. M. Van den Tempel,
  Unilever Research,
    Vlaardingen, The Netherlands.

Professor K. Walters,
  Department of Applied Mathematics,
    University College of Wales, Aberystwyth, U.K.

Professor J.L. White,
  Polymer Engineering,
    The University of Tennessee,
      Knoxville, U.S.A.

LIST OF CONTENTS

Page

CHAPTER 1

FUNDAMENTAL CONCEPTS

By

Ken Walters

CONTENTS

1. INTRODUCTION

In November 1978, the Belgian Group of Rheology organised a three-day "Rheometry Workshop" at the Catholic
University of Leuven. The workshop dealt with the
rheometry of fluids, including polymer melts and solutions, suspensions, rubber compounds and semi-solids.
The objective of the meeting was to provide an extensive survey of current rheometrical techniques and
available commercial equipment. Arrangements were
accordingly made with the manufacturers of viscometers
and rheometers to have their instruments available at
the workshop. More than twenty different instruments
were exhibited and the attendance at the workshop
reached 200, facts which clearly proclaim the continued relevance of Rheometry. Industrial participation
at the workshop was substantial, indicating the growing awareness of the practical importance of the
subject.

These and related observations provide sufficient
justification for an attempt to compile an up-to-date
book on industrial aspects of Rheometry, especially
since fundamental work on basic Rheometry has tended
to ignore the materials and conditions met in many
industrial processes. It is common practice for research into fundamental aspects of the subject to be
limited by the following convenience factors:

(i) experiments are usually carried out at room temperature.

(ii) polymer solutions dominate experimental pro-grammes, with aqueous solutions of polyacrylamide (or polyox) and polyisobutylene solutions being the pre-ferred test fluids.

(iii) careful mixing ensures that the test fluids can be considered to be homogeneous. They are also invariably isotropic.

(iv) "difficult" time-dependent effects associated with thixotropy and anti-thixotropy are avoided if at all possible and even yield stresses are unpopular in most fundamental studies.

Factors (i) to (iv) can be defended without diffi-culty in any fundamental rheometry study, but indus-trialists who have to handle complex materials with some or all of the difficult problems associated with temperature variations, suspended particulate matter, yield stresses, structural breakdown and time-dependent effects need convincing of the value of fundamental studies in the resolution of their peculiar problems. One of the objects of the present book is to address itself to this important problem.

In this chapter, we prepare the ground by outlining some of the basic concepts of rheology and rheometry, including the ideas of solid-like and fluid-like be-haviour and the introduction of the fundamental mater-ial functions associated with rheometrical experiments.

2. SOLIDS, LIQUIDS AND THE DEBORAH NUMBER

In 1678, Robert Hooke developed his "True Theory of Elasticity". He proposed that "the power of any spring is in the same proportion to the tension thereof". Very simply, if you double the tension in a spring, you double the extension. This forms the basis for Classical Elasticity Theory of solid mechanics.

At the other end of the spectrum, Isaac Newton gave attention to liquids and in the Principia published in 1687 there is the following hypothesis: "The resis-tance which arises from the lack of slipperiness of the parts of the liquid, other things being equal, is proportional to the velocity with which the parts of the liquid are separated from one another." (see Figure 1). This lack of slipperiness is what we now call "viscosity". This is synonymous with internal friction and is a measure of resistance to flow. The motion in Figure 1 is called steady simple shear flow. The force per unit area required to produce the motion (i.e. the shear stress) is denoted by τ and is proportional to the velocity gradient U/d, the constant of proportionality η being the coefficient of viscosity.

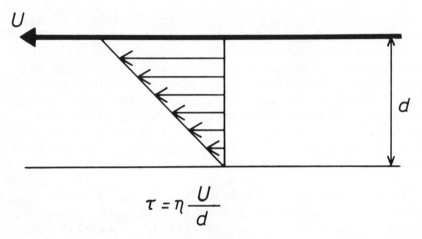

$$\tau = \eta \frac{U}{d}$$

Fig. 1 Newton's postulate.

Glycerine and water are common liquids that obey Newton's postulate. For glycerine the viscosity in SI units is of the order of 1 Pa.s, whereas the viscosity of water is about 1 mPa.s, i.e. one thousand times less viscous.

So far, we have discussed "solids" and "liquids" without defining what we mean by these terms. Often, it is taken for granted that when we talk or write about a solid or a liquid, the hearer or reader knows exactly what we mean. If we asked the layman for definitions, we would get a variety of answers, but they would all contain the same basic ideas: "A liquid is a material that flows under its own weight, that finds its own level when placed in a beaker." "A solid is a material that suffers from homesickness - when we stretch a rubber band it has a strong desire to return to its original configuration; when we drop it on the floor, it bounces - it has elastic properties."

These would be acceptable answers and the layman would not expect any ambiguity to arise from the respective definitions. This is, however, an oversimplification of what is a complex situation. Take, for example, the case of the silicone commonly called bouncing (or silly) putty, which is sold as a children's toy, but is also employed in physiotherapy and is used for packing electronic equipment. On the basis of popular definitions of solid-like and fluid-like behaviour, the silicone is a liquid, albeit a very viscous one. It certainly finds its own level when placed in a container - given sufficient time. But this is not the whole story as its nickname "bouncing putty" would indicate. Indeed, as a toy it is sold as "the real solid liquid" and this does not violate the Trade Descriptions Act. Here, we have a liquid which possesses some of the properties we usually associate with solids. We call it an elastico-viscous liquid or simply an elastic liquid.

There are also solid-like materials which will flow for a time. They are called viscoelastic solids.

We are driven to an obvious conclusion. Materials such as water and rubber are at opposite ends of a whole spectrum of possible material behaviours. There is, in fact, no clear line of demarcation between solids and liquids.

At this stage, it is helpful if we consider again some simple tests which can be carried out on bouncing putty. In a slow-flow process, the material behaves like a liquid, e.g. the putty slowly finds its own level in a container. On the other hand, in the faster time process as the putty is dropped and strikes the ground, it bounces - which is a solid characteristic. So a given material can behave like a solid or a liquid depending on the time scale of the deformation process.

Professor M. Reiner, one of the founders of the modern science of Rheology, appealing to a quotation from the 5th Chapter of the Book of Judges which reads "The mountains melt (or flow) before the Lord" introduced the helpful concept of a "Deborah Number" D_e, defined by

$$D_e = \frac{\lambda}{t} , \tag{1}$$

where t is a characteristic time of the flow process and λ is a characteristic time of the material. This would be infinite for an ideal elastic solid and, in principle, zero for a Newtonian viscous liquid. In fact, for water, λ is typically 10^{-13} sec. which is near enough to zero for most purposes. For lubricating oils, λ is of the order of 10^{-5} sec. and for the polymer melts used in the plastics processing industries, the characteristic time can be as high as a few seconds.

High D_e corresponds to solid-like behaviour and low D_e to liquid-like behaviour. A material can appear solid-like either because it is, i.e. it has an infinite characteristic time or because the deformation process it is exposed to is very fast. One consequence of this is that even mobile liquid systems with a low characteristic time λ can behave like elastic solids in a very fast flow process.

Having in one sense confused the situation, we now attempt unambiguous definitions for a solid and a liquid. We define a solid as a material that will not continually change its shape when subjected to stresses (i.e. for a given stress field there will be a fixed deformation, which may or may not be reached instantaneously). A liquid is a material that will change its shape continually (i.e. it will flow) when subjected to stresses irrespective of how small these stresses may be.

It is well known that some materials behave as solids under small stresses and as fluids after a critical value of the stress (called the yield stress) has been exceeded. These plastic materials are solids under our classification scheme.

Our main emphasis in this book is on mobile liquid systems, but we shall find it convenient to include in our discussions materials with a yield stress.

3. STEADY SIMPLE SHEAR FLOW

By far the most popular rheometrical flow is that usually referred to as steady simple shear flow. Such a flow can be generated in various geometries [2] but we shall confine attention to a simple rectangular Cartesian representation with velocity components

$$v_1 = qx_2, \qquad v_2 = v_3 = 0, \qquad (2)$$

where q is a constant velocity gradient or shear rate.

The corresponding stress distribution can be written in the form [2]

$$\left.\begin{aligned} P_{12} &= \tau(q) = q\eta(q), \\ P_{11} - P_{22} &= \nu_1(q), \\ P_{22} - P_{33} &= \nu_2(q), \end{aligned}\right\} \qquad (3)$$

where η is called the apparent viscosity (or shear-dependent viscosity), ν_1 is the first normal stress difference and ν_2 is the second normal stress difference. For a Newtonian liquid, η is a constant and ν_1 and ν_2 are zero. For an elastic liquid without a yield stress, we must have

$$\eta(q) \to \eta_0, \quad \nu_1(q) \to 0, \quad \nu_2(q) \to 0 \quad \text{as } q \to 0, \quad (4)$$

where η_0 is a constant known as the limiting viscosity at small rates of shear. (4) implies that all elastic liquids will behave as Newtonian liquids in the limit

of vanishingly small shear rates. Normal stresses
manifest themselves when terms of order q^2 are non-
negligible, and the shear stress τ ceases to be a
linear function of q when terms of order q^3 are
important.

3.1 The Shear Stress

There is no doubt that from a practical standpoint
the shear stress τ (or equivalently the associated
apparent viscosity η) is the most important rheometrical function. τ(or η) determines the pumping requirements in most transport processes and is also of cru-
cial importance in lubrication studies. Indeed, the
practical sections of this book are dominated by discussions of the apparent viscosity.

Representative (τ,q) rheograms are shown in Figure 2.
For a Newtonian liquid, there is a linear relation
between τ and q. Some (plastic) materials will not
flow until a critical yield stress τ_y is exceeded. If
the (τ,q) graph for $\tau > \tau_y$ is still a straight line,
we have what is commonly referred to as a Bingham
plastic material.

Most rheologically complex materials without a yield
stress show the type of shear-thinning behaviour illustrated in Figure 2 and again in Figure 3. Such behaviour is sometimes called "pseudo-plastic" from an obvious comparison with typical plastic behaviour, but
it is now more commonly referred to as shear-thinning.

We see in Figure 3 that the viscosity falls monotonically from the zero shear viscosity η_0 (in the 'first
Newtonian region') to a lower value η_s (in the 'second
Newtonian region'). The difference between η_0 and η_s
can be substantial. For example, it is possible for a
liquid to have the consistency of glycerol at very low
shear rates and that of water in the second Newtonian
region, indicating a fall in viscosity of three orders
of magnitude.

It is often found in practice that the region of
falling viscosity is very well approximated by a

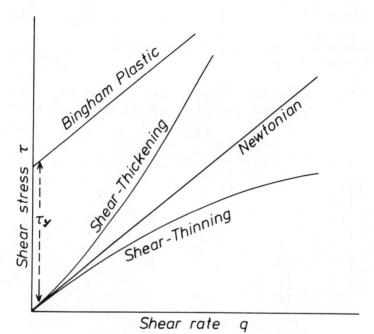

Fig.2 Representative (τ,q) rheograms.

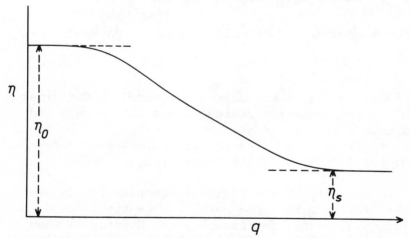

Fig.3 Schematic diagram of typical shear-thinning behaviour.

"power-law" equation of the form

$$\eta(q) = \kappa q^{n-1} , \quad n \leqslant 1 , \tag{5}$$

where n and κ are constant parameters. Clearly n = 1 corresponds to a Newtonian liquid.

Figure 4 shows power-law behaviour for one of the popular test fluids used in fundamental studies.

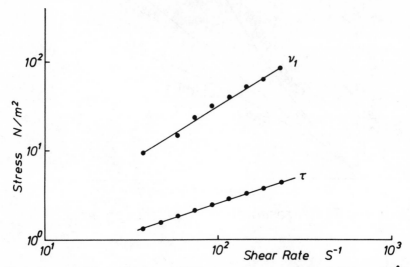

Fig.4 Viscometric data for a 0·5% aqueous solution of Polyox. 23°C.

Equation (5) is not able to accommodate the first and second Newtonian regions shown in Figure 3 and various ingenious adaptations to (5) are to be found in the literature. Readers are referred to references [3] and [4] for a fuller discussion.

The majority of rheologically complex fluids are shear-thinning but some notable exceptions (sand and clay suspensions, for example, cf. Chapter 7) show the alternative shear-thickening behaviour illustrated in Figure 2. Such behaviour, represented by n > 1 in (5), is sometimes called dilatancy, but this terminology is deprecated [1].

Finally, we note that materials with a yield stress can also show shear-thinning or shear-thickening behaviour above the yield stress as well as the linear behaviour illustrated in Figure 2 for the classical Bingham plastic material.

Rigorous measurements of the apparent viscosity $\eta(q)$ are usually carried out either in a capillary (Poiseuille) flow or in rotatory devices of the cone-and-plate type. Historically, Poiseuille flow has played a dominant role and is certainly one of the easiest rheometrical measurements to perform. However, the interpretation of experimental results is not completely straightforward [2]. Rotary instruments usually require sophisticated initial alignment, but when this difficulty is overcome, the interpretation of experimental results is very easy. For example, for the cone-and-plate flow shown schematically in Figure 5, the relationship between the torque C on the wire supporting the plate and the rotational speed Ω_1 of the cone is given by

$$C = \frac{2\pi a^3}{3} q\eta(q) \, , \qquad q = \frac{\Omega_1}{\theta_0} \, , \qquad (6)$$

where a is the radius of the cone and θ_0 is the gap angle.

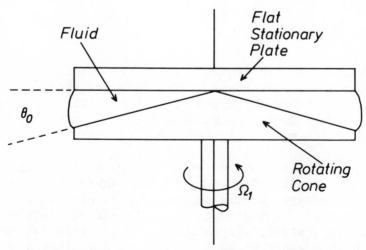

Fig.5 Basic cone and plate geometry

The Brookfield device, in which a spindle is rotated in a bath of test fluid and the couple on the spindle measured (cf. Chapter 7), is very easy to use but the interpretation of the experimental results in terms of (η, q) data is far from straightforward. However, recent research into such devices [5] offers the hope that the conversion from (C, Ω) to (η, q) data can be made without too much mathematical computation.

Other relatively simple viscometer devices such as the Ford cup are also mentioned in the course of this book (cf. Chapter 2). We await detailed research studies on the optimum way of interpreting the experimantal results in terms of fundamental rheometrical parameters or functions.

3.2 The First Normal Stress Difference

When researchers are seeking a quantitative measure of elasticity in a rheologically-complex fluid they usually turn to the first normal stress difference ν_1. This function can now be obtained with relative ease from a variety of commercial rheometers, the preferred geometry in most of them being the cone-and-plate set-up shown in Figure 5. The first normal stress difference ν_1 is related to the total normal force F on the plate through the simple formula

$$F = \frac{\pi a^2}{2} \nu_1 . \tag{7}$$

In highly elastic liquids, it is not unusual for ν_1 to be orders of magnitude bigger than the shear stress τ. This is shown in Figure 4, which further suggests that a power-law representation for ν_1 is also a good approximation over a limited range of shear rates.

Except for the limiting case of very small shear rates, it is generally conceded that shear-thinning behaviour and substantial normal-stress effects are characteristic features of the behaviour of elastic liquids in a steady simple shear flow. Practical experience with a range of such liquids soon indicates that the two manifestations of non-Newtonian behaviour are related to each other in the sense that the greater the normal stress levels the greater is the drop in viscosity. One notable exception is the so-called Boger liquid (a polyacrylamide solution in a maltose syrup/water mixture [6]) for which the viscosity is found to be sensibly constant over a wide range of shear rates with the corresponding normal stress levels nevertheless substantial (see Figure 6 where the slope of the shear stress curve is 1 and that of the first normal stress difference is 2). Such behaviour does not conflict with any fundamental principle of continuum mechanics and is not unusual in that sense, but it is useful to have available a test fluid which shows measurable normal stresses and a constant viscosity over a non-trivial shear-rate range.

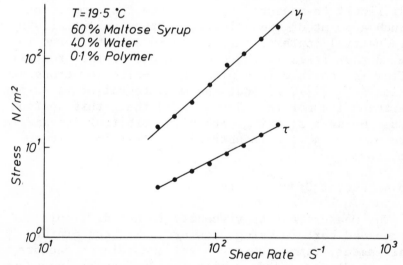

Fig. 6 Viscometric data for a Boger fluid.

The measurement of normal stress levels in materials with a yield stress is difficult, especially when instruments employing a servo-controlled normal-force technique are used. Recent tests with the Torsional Balance Rheometer have alleviated some of the relevant problems (cf. Chapter 3).

3.3 The Second Normal Stress Difference

During the 1960's and early 70's, the measurement of the second normal stress difference occupied the attention of many research groups working in fundamental rheometry and also some from industrial establishments (see, for example, [2]). In general terms, experimental results were characterized by contradictions, significant scatter and uncertainty. However, a careful scrutiny of the more reputable measurements has led to a consensus that ν_2 is usually much smaller than ν_1 (with $|\nu_2| < 0.2|\nu_1|$) and also of opposite sign. Exceptions to this are to be found in the literature and it must not be assumed that this consensus is sacrosanct and has to apply to all systems.

It is the relative smallness of ν_2 that results in some of the difficulties encountered in its measurement [2]. Indeed, it has been said that if ν_2 is so difficult to measure, it cannot be that important! Such a point of view clearly carries weight within an industrial context, but it should also be borne in mind that it is ν_2 which determines whether rectilinear flow is possible in a pipe of non-circular cross section (cf. [7]), so that the measurement of ν_2 is not without its merits. Having said that, this section must be seen as being within the orbit of fundamental rheometry and the industrial interest in ν_2 is now minimal.

3.4 Recent Developments

The measurement of viscosity is not difficult and provided certain safeguards are taken concerning alignment, temperature control and related factors, can be determined to a high level of precision. Current interest now centres on the possibility of making meaningful (η, q) measurements from such convenient shop-floor instruments as the Brookfield viscometer and the Ford-cup arrangement. Clearly these studies will be important within an industrial framework.

The determination of ν_1 for modest shear-rate ranges is now a well established and comparatively simple task and the last ten years has seen the proliferation of sophisticated rheometers designed in part to measure ν_1. These instruments usually find their way into industrial research laboratories but are not often located on the shop floor.

The availability of flush mounted pressure transducers has resulted in the introduction of two new commercial rheometers[#] which make use of "hole pressures" and "exit pressures", respectively. The first, developed by Lodge, covers a wide range of possibilities and in one adaptation it is hoped to be able to reach shear rates of the order of 10^6 sec^{-1} (A.S. Lodge, private communication).

It is the extension of existing shear-rate ranges to accommodate the very high shear rates encountered in many industrial situations which has recently led to some interesting new developments. The conventional jet-thrust technique is now thought to be suspect due to the unknown conditions at the capillary exit [8] and attention has turned to torsional flow to provide the required data [9]. The associated Torsional Balance differs from the conventional torsional-flow rheometer in that it is the normal load which is fixed in the T.B. and the measured gap between the plates gives the associated rim shear rate. The T.B. gives the normal stress combination $\nu_1 - \nu_2$ rather than ν_1 or ν_2 separately but at the high shear rates involved ($\sim 10^5$ sec^{-1}) one is basically interested in normal stress levels so that the rheometer provides meaningful information, especially in view of our earlier comments concerning the relative size of ν_2.

Now that the jet-thrust technique is thought to be suspect[*], it would be interesting in future to compare

[#] The Seiscor/Lodge Stressmeter and the Seiscor/Han Rheometer.

[*] This suspicion refers to the quantitative determination of meaningful normal stress functions. It does not imply that the technique cannot supply a useful measure of "elasticity" at high shear rates.

results from the T.B. and Lodge's new high shear-rate
rheometer.

At the present time, work associated with the determination of ν_2 has to be of very high quality to attract the attention of other researchers, and the number
of papers on the subject has fallen off dramatically.
However, we know of at least two noteworthy recent
studies from two respected workers in the field
(Lodge [10] and Tanner [11]). Of significant interest
to researchers in fundamental rheometry will be
Tanner's conclusion [11] that the Boger liquid has a
zero second normal stress difference (to within
experimental error).

A further subject of current interest connected with
shear flows (although not steady shear flows) concerns
the so-called "jump" tests. In these tests, measurements of stress (and sometimes strain when a jump in
stress is involved) are made following an imposed jump
(or step) in shear strain or shear strain rate. Stress
relaxation tests and the popular stress overshoot
experiments are popular examples of jump tests (see,
for example, [12,13]). Recent interest has however
shifted to tests involving step changes in strain.
For a jump shear strain s_0 at time $t = 0$ in a liquid
otherwise at rest, interest centres in the ratio

$$\frac{\nu_1(t)}{s_0 \tau(t)} \quad (t > 0) \; .$$

Various rheological models predict [14] this ratio to
be either

  (i)   1
 (ii)   a constant other than 1
(iii)   a function of s_0
 (iv)   a function of s_0 and t
  (v)   indeterminate.

A simple jump test of the kind envisaged by Lodge
can thus eliminate large classes of *a priori* acceptable equations for describing a given test material.
This is an important area of current research in

fundamental rheometry.

4. OSCILLATORY SHEAR FLOW

Here, we are concerned with a small-amplitude oscillatory shear flow, which for convenience we represent by the rectangular Cartesian velocity field

$$v_1 = \varepsilon \omega x_2 e^{i\omega t}, \quad v_2 = v_3 = 0, \tag{8}$$

where ε is small enough for second and higher order terms to be neglected. In (8), $i = \sqrt{-1}$ and the real part is implied. The corresponding stress field can be written in the form [2]

$$P_{12} = \eta^*(\omega) \varepsilon \omega e^{i\omega t}, \quad P_{11} - P_{22} = P_{22} - P_{33} = 0, \tag{9}$$

where η^* is the complex dynamic viscosity. It is usual to express η^* in the form

$$\eta^* = \eta' - i \frac{G'}{\omega}, \tag{10}$$

where $\eta'(\omega)$ is the dynamic viscosity (not to be confused with the dynamic viscosity of classical fluid mechanics) and G' is the dynamic rigidity. For completeness, we note that it is sometimes customary to write (cf. Chapter 4)

$$\eta'' = G'/\omega, \tag{11}$$

$$G'' = \omega\eta'. \tag{12}$$

For illustration purposes, we give in Figure 7 complex viscosity data for a Boger liquid. The monotonic decrease in η' with frequency and the corresponding rise in G' are characteristic features of the behaviour found for mobile elastic liquid systems.

Experimenters now have an embarrassment of choice concerning the practical measurement of η^*. They can either use an "unsteady" shear flow similar to the one described by (8) and characteristic of instruments like the Weissenberg Rheogoniometer, or alternatively they can employ a "steady" flow of the type generated in the Maxwell Orthogonal Rheometer or Képés Balance Rheometer. The pros and cons of the alternative measurements have been fully discussed in [2] and there is now a number of commercial instruments available for carrying out both types of measurement.

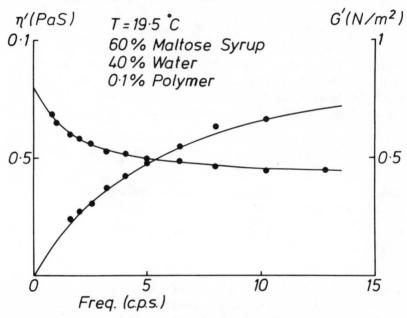

Fig.7 Dynamic data for a Boger fluid.

Materials with a yield stress have sometimes been studied under conditions of small-amplitude oscillatory shear flow. The type of response in one of the conventional instruments which uses an input and output oscillatory response is shown schematically in Figure 8, the flattening of the output trace (the amount of which depends on the input amplitude) being a characteristic feature of observed behaviour if the strain-rate sweep takes the material outside its solid-like regime. Clearly a harmonic study of the output response using a transfer function analyser or similar instrument is of little application other than for simply demonstrating the existence of a yield stress.

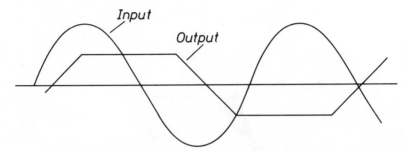

Fig.8 Schematic representation of input and output traces for a material with a yield stress in a small amplitude oscillatory shear flow.

Commercial instruments now exist for subjecting test fluids to a combined steady and oscillatory shear of the form

$$v_1 = qx_2 + \varepsilon\omega x_2 e^{i\omega t}, \quad v_2 = v_3 = 0 \ , \qquad (13)$$

where q is a constant. It is still possible to represent the shear stress in terms of a complex viscosity which is now a function of q as well as ω. A schematic representation of elastico-viscous behaviour in such a flow is shown in Figure 9.

The halcyon days of combined steady and oscillatory shear testing in fundamental studies are now passed, but it is interesting to note that such testing has found a practical application in at least one industry (Chapter 4).

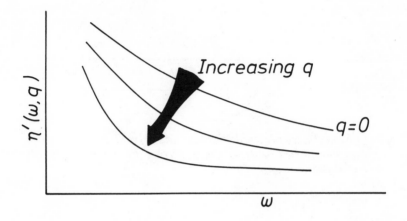

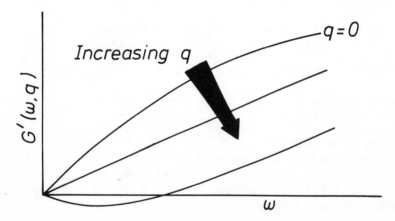

Fig.9 Schematic representation of dynamic data in a combined steady and oscillatory shear flow.

5. EXTENSIONAL FLOW

While steady simple shear flow and oscillatory shear flow generally occupied the attention of rheologists in the 1960's, the 1970's have been characterized by an expanding interest in extensional (or elongational) flows, with a proliferation of research papers on the subject, the slow emergence of commercial extensional viscometers and the growing awareness of the practical importance of extensional flows in many industrial processes. A useful and up-to-date summary of theoretical and experimental aspects of the subject is contained in a recent book by Petrie [15].

The author's own fascination with the subject was heightened recently whilst making a film on certain aspects of rheologically complex behaviour with a colleague, Dr. J.M. Broadbent. In an attempt to demonstrate shear-thinning behaviour in dilute polymer solutions we used two syringes[#], one filled with an aqueous solution of polyethylene oxide and the other with a glycerol/water mixture, the ratio of glycerol to water being chosen to ensure that the fluids in the two syringes drained at the same rate while falling under gravity. The top of the syringes were then pressurized with the expectation that the dilute polymer solution would now drain more readily on account of the shear-thinning associated with the higher shear rate in the Poiseuille flow on either side of the syringe contraction. To our surprise and embarrassment, the Newtonian glycerol/water mixture drained more readily. A moment's reflection was sufficient to indicate that the extensional stresses associated with the converging flow regime at the sudden contraction were a dominant and overriding influence in deciding the resulting pressure drop/flow rate behaviour. Extensional viscosity considerations were more important in our experiment than the shear viscosity and there was a clear indication that the extensional viscosity

---

[#] Which happened to be to hand at the time. Each syringe consisted essentially of two relatively long capillaries of different diameter with a sudden contraction between them.

of the polymer solution was significantly greater than that for the Newtonian liquid.

In discussing extensional flows, it is convenient to confine consideration to the rectangular Cartesian flow field (see also Chapter 5, section 3).

$$v_1 = kx_1, \quad v_2 = -\frac{k}{2}x_2, \quad v_3 = -\frac{k}{2}x_3, \quad (14)$$

where k is the constant extensional strain rate. The corresponding stress distribution can be written in the form

$$p_{11} - p_{22} = p_{11} - p_{33} = k\,\eta_E\,(k)\ , \quad (15)$$

where η_E is the extensional (or elongational) viscosity. For a rheologically complex fluid, the ratio of extensional viscosity to shear viscosity is defined as the Trouton ratio. In the case of a Newtonian liquid, this ratio is 3.

Some workers in the field do not believe that we have yet reached a position where we can be categorical about the extensional behaviour of rheologically complex fluids. This scepticism is due in part to the difficulty of making accurate measurements of η_E over a sufficiently wide range of extensional strain rates k. The position should be eased in future by the emergence of commercial extensional rheometers and also by the development of the related background theoretical work to ensure that these and other rheometers do in fact give the data they purport to provide.

In the meantime, a consensus of opinion would support the proposition that within this area of research we must be careful to distinguish between dilute polymer solutions on the one hand and concentrated polymer solutions and polymer melts on the other. For the former, there is sufficient theoretical evidence from microrheological studies and also experimental support (often of a qualitative nature) for the belief that extensional viscosities can be several orders of magnitude higher than the corresponding (shear) apparent viscosity. The simple experiment described earlier would lend support to this point of view.

From these observations and the fundamental requirement that $\eta_E(k) \to 3\eta(q)$ as k and q tend to zero, we are led to the conclusion that, for dilute polymer solutions at least, η_E must be a rapidly increasing function of k.

Existing experimental work on concentrated polymer solutions and polymer melts sometimes indicates that η_E increases with k, sometimes that it decreases, while at other times a small maximum is possible [2]. There is no fundamental reason why all these possibilities should not apply to different fluid systems, but there is certainly need for more detailed studies to clarify the situation. One thing does appear to be incontrovertible and that is that the dependence of η_E on k for concentrated solutions and melts is relatively weak. However, the resulting Trouton ratios can still be substantially greater than 3 on account of the shear-thinning behaviour associated with the apparent viscosity $\eta(q)$.

The search for extensional viscometers which permit easy experimentation and meaningful interpretation of the data goes on. Work on lubricated dies, converging flows and spinning experiments, amongst others, is in progress at the present time in many laboratories. The main difficulty would seem to be the correct interpretation of data from experiments which are relatively easy to perform (e.g. spinning and converging flows) and the practical construction, if possible, of those experiments which are fairly easy to interpret.

The precise interpretation of motions with a high extensional component in terms of the material function $\eta_E(k)$, which is an important pursuit to those interested in fundamental rheometry, may seem of little consequence to the industrialist who simply wants some "measure" of a material's resistance to extensional stresses in motions of relevance to him. It may well be that existing instruments, commercial or otherwise, will be sufficient for his purpose. For example, an

industrialist interested in fibre spinning may find
elaborate treatises and discussions about the correct
way of interpreting data from Ferguson's spin-line
rheometer [16] to be largely irrelevant, since all he
probably needs is an average measure of stress and
rate of deformation in a process clearly of relevance
to him.

6. THIXOTROPY AND ANTI-THIXOTROPY

In addition to the change in material properties
which can be brought about by changes in the time
scale of a flow process and by the strength of that
process, there are also changes which can be brought
about in some materials by simply waiting long enough.
The stirring (shearing) of some materials at a constant rate can result in a substantial lowering of the
viscosity with time, with a gradual return to the
initial viscosity when the stirring is stopped. This
effect is called "thixotropy" and has been actively
encouraged for example in the so-called non-drip
paints. It is also a well-known phenomenon in sauces
and ketchups. We all feel for the embarrassed diner
who cried

"When I shake the ketchup bottle,
None'll come or else the lot'll!"

Thixotropy is not an easy phenomenon to study either
from a theoretical or an experimental standpoint. This
may account for the relatively small number of papers
on what is an important practical problem. Many of
the papers which have appeared on the subject are
covered in the recent review by Mewis [17].

The recent advent of commercial rheometers with a
programmable drive unit should facilitate future experimental studies of time-dependent phenomena, but there
is no sign as yet of significant research effort on
theoretical aspects of the problem.

Some time-dependent materials show an increase in
apparent viscosity with time at a fixed shear rate
with a gradual recovery when the stress is removed.
This phenomenon is known by one of the three terms:

anti-thixotropy, negative thixotropy, rheopexy. An
example of such behaviour in an alkaline perbunan latex
is illustrated in Figure 10. The fluid returned to
its equilibrium state after about 2 minutes from the
taking of the second picture.

7. RESEARCH LABORATORY versus SHOP-FLOOR ACTIVITY

Scientists interested in industrial rheometry fall
naturally into three categories:

(i) There are those from academic establishments who
are often interested in fundamental rheometry, but who
have been forced by economic pressures (related to
research-funding problems) to make their work industrially relevant.
(ii) Secondly, the major industries usually possess
research laboratories, employing researchers having an
interest in rheometry. Such laboratories invariably
possess one or more of the "jumbo" rheometers – the
Weissenberg Rheogoniometer, the Mechanical Spectrometer
or the new Instron Rheometer – and carry out detailed
rheometrical measurements on the materials of relevance
to their employers.
(iii) Finally, there are the shop-floor workers who
usually possess unsophisticated rheometers and who do
not have the time, facilities (or, sometimes, the inclination) to make more than single-point measurements.

Whereas contact between scientists in categories (i)
and (ii) has been (and is) extensive and fruitful,
those in category (iii) are isolated. The situation
is amply illustrated by the words of Dr. Beazley in
his chapter on industrial suspensions: "Industrial
control, however, tends to ignore the complex rheology
shown up by research, treats its materials as rheologically simple and uses simple – in some cases primitive
– methods for rheological control." It is clear from
other chapters in this book that Beazley's comments
could apply equally well to other industries. Furthermore, to change practices at the shop floor level is
a very slow process and new standards and tests are
not introduced without significant effort. If sophisticated rheometry has a part to play in controlling

A   B

Fig. 10 An alkaline perbunan latex before (A) and after (B) vigorous shaking. (cf. Reference [18]).

industrial processes as those in categories (i) and
(ii) would advocate, the scientists on the shop floor
are clearly unimpressed. Here, conservatism (some
would say, pragmatism,) rules.

It would be wrong, however, to put the blame for
"the great divide" solely at the feet of the shop-
floor worker. He has a right to be unimpressed by his
colleagues from the research laboratory so long as
they provide data which are of no relevance to the
processing conditions met in practice. These may
involve very high shear rates, elevated temperatures,
short time scales, high pressures and other headaches.
Even the jumbo rheometers find difficulty working in
these rarified conditions and, in future, one might
hopefully see the emergence of more industrial rheometers, custom built to meet the conditions experienced
in the relevant processes.

That a gulf exists between research laboratory and
shop-floor rheometry is indisputable and is made
abundantly clear in the remaining chapters of this
book. Hopefully, elucidating the problem will generate
genuine efforts to bridge the gap, to the ultimate
benefit of all concerned.

REFERENCES : PREFACE AND CHAPTER 1

[1] B.S.I. Glossary of Rheological Terms BS5168 : 1975

[2] K. Walters, Rheometry, Chapman and Hall, 1975

[3] R.B. Bird, R.C. Armstrong and O. Hassager,
Dynamics of Polymeric Liquids : Vol. 1 Fluid
Mechanics, John Wiley and Sons, 1977

[4] W.R. Schowalter, Mechanics of Non-Newtonian
Fluids, Pergamon, 1978

[5] R.W. Williams, Rheol. Acta 18 (1979) 345

[6] D.V. Boger, J. Non-Newtonian Fluid Mechanics 3
(1977/78) 87

[7] P. Townsend, K. Walters and W.M. Waterhouse,
J. Non-Newtonian Fluid Mechanics 1 (1976) 107

[8] J.M. Davies, J.F. Hutton and K. Walters, J. Non-
Newtonian Fluid Mechanics 3 (1977/78) 141

[9] D.M. Binding and K. Walters, J. Non-Newtonian
Fluid Mechanics 1 (1976) 277

[10] A.S. Lodge, Private communication

[11] M. Keentok, A.G. Georgescu, A.A. Sherwood and
R.I. Tanner, To appear in J. Non-Newtonian
Fluid Mechanics

[12] M.A. Lockyer and K. Walters, Rheol. Acta 15
(1976) 179

[13] P. Attané, P. Le Roy and G. Turrel, To appear in
J. Non-Newtonian Fluid Mechanics

[14] A.S. Lodge, To appear in J. Non-Newtonian Fluid
Mechanics

[15] C.J.S. Petrie, Elongation Flows, Pitman, 1979

[16] J. Ferguson and N.E. Hudson, J. Phys.(E) $\underline{8}$ (1975) 265

[17] J. Mewis, J. Non-Newtonian Fluid Mechanics $\underline{6}$ (1979) 1

[18] D.C-H. Cheng, Nature, $\underline{245}$ (1973) 93

CHAPTER 2

DETERGENTS

By

H.A. Barnes

CONTENTS

1. INTRODUCTION

Over the past 30 years there has been a general trend in the detergent and allied industries towards liquid-like cleaning products (making the measurement of flow properties a very important study). Originally solid soap and scouring powders were the only products available. These were supplemented in the mid-1920's by soap powders, then in the '40's by non-soap detergent (NSD) powders. However from then on the trend in new products has been towards liquids; for dishwashing (1958), hard surface cleaning liquid scourers for floors and walls (1960), fabric softening (1967), thickened bleaches (1969) and liquid abrasive cleaners (1970). In the U.S.A. a current trend is towards liquid detergents for washing fabrics, although this is not yet the case in Europe.

Personal washing products have also tended to move to liquid-like products with hair shampoos and conditioners, shower gels and body shampoos, toothpaste, skin cleaners and so on.

The rheology of these liquid-like products is either dictated by consumer preference (e.g. the expectation of a concentrated liquid to be viscous and not watery) or by the need to impart some technical benefit (e.g. liquid abrasive cleaners need a fast rebuilding yield

34

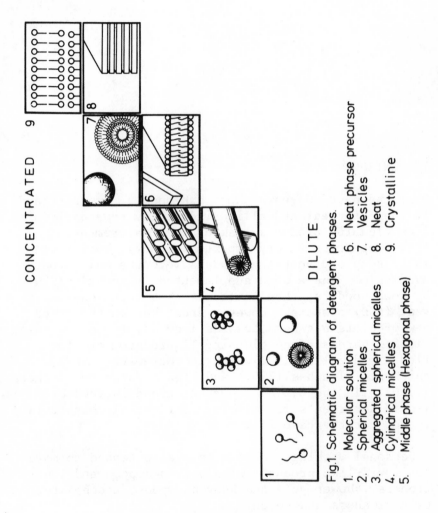

Fig.1. Schematic diagram of detergent phases.

1. Molecular solution
2. Spherical micelles
3. Aggregated spherical micelles
4. Cylindrical micelles
5. Middle phase (Hexagonal phase)
6. Neat phase precursor
7. Vesicles
8. Neat
9. Crystalline

stress to suspend abrasive particles). The role of rheometry – the measuring arm of rheology – is to quantify these rheological parameters of 'thickness', 'suspendability', etc.

The particular problems encountered in the rheology of detergent-type liquid products stem from the basic detergent molecules and their predisposition to associate into ordered multiple units. Detergent molecules have a number of different atomic formulae but they all have one thing in common – a hydrophilic (water-loving) headgroup and a hydrophobic (water-hating) hydrocarbon 'tail'. These molecular properties lead to the formation of various structured phases, as shown in Fig. 1, when the detergent is dispersed in water*.

These structures affect the rheology, and over the years it has been one task of the industry to learn how to manipulate these phases to produce desired rheological properties.

Rheometry is used in the industry at various levels of sophistication to provide information for different purposes:

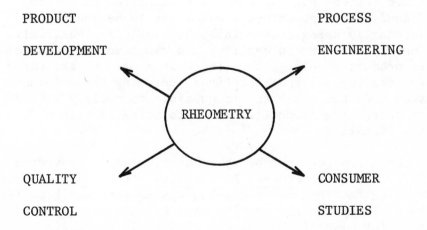

PRODUCT

DEVELOPMENT

PROCESS

ENGINEERING

RHEOMETRY

QUALITY

CONTROL

CONSUMER

STUDIES

* These phases become 'inverted' in oil but produce analogous phases (cf. Section 3.1).

Rheometry can of course be performed over a wide range of deformation regimes and rates, but often information is required over only a limited range, and the rheologist therefore needs to know the end use of the information he is providing to avoid overspecification and confusion.

Dealing with the four common purposes above in turn we can first state that the aim of Quality Control rheometry is to ensure that the product leaving the factory is within the rheological product specification. This task is often performed at the manufacturing stage, where the plant operator performs a simple rheometry test; the skill, time and resources available to him are limited, so it is important that a simple, cheap, yet relevant, rheometry test is specified.

In Product Development the aim of rheometry is twofold – first to help indicate (via rheological manifestations) what the microstructure is in various prototype products and so to manipulate the rheology into a desired range via calculated changes in the microstructure. This generally calls for the greatest sophistication in rheometrical techniques e.g. the kind of information available from a rheogoniometer, and invariably requires additional information from other sources e.g. spectroscopy, electron microscopy. Secondly, the prototype product has to be manufactured eventually using commercially available raw materials. These often vary in quality, and can lead to deviations in product rheology, thus there is a need to set careful rheological specifications (defining the deformation mode and rate) and if possible to specify methods of correcting product rheology to bring it within specification.

Rheometry need not be so sophisticated in providing information for Process Engineering. Usually viscosity data in the form of "power-law" parameters are required to scale-up mixing, pumping etc. from laboratory ($\sim$1 kg) to pilot plant ($\sim$500 kg) and full-scale production ($\sim$10,000 kg).

For Consumer Studies the problem generally is to assign the correct mode and rate of deformation to simulate the way in which the consumer judges the product – does he/she judge by pouring, shaking, watching it spread, rubbing it into the hands and so on? Once a decision has been made as to the correct mode and rate of deformation, preferred rheological properties can be found from consumer testing. A range of possible products can be screened quickly in the laboratory using the correct rheometrical conditions which simulate the consumer situation. This method is preferred because of the time-consuming and expensive nature of consumer testing.

The total need of an industry for rheological material parameters thus requires a detailed knowledge of the end use of the information e.g. a knowledge of the currently available structural models that connect particle interactions/sizes with yield stress, shear and extensional viscosity, elasticity etc. and the available correlations of mixing with Reynolds number (and possibly Weissenberg and Deborah numbers). It also requires a good understanding of rheometry so that the modes and rates of deformation of particular instruments can be properly defined. This latter aspect will be considered in the next section.

2. RHEOMETRIC TECHNIQUES

2.1 Introduction

Table 1 shows in diagrammatic form the range of products to be considered in terms of their consistency (water-thin to soft solid) and their complexity (Newtonian to very non-Newtonian with possible time-dependence).

Obviously to provide a complete characterisation for all these materials would need a wide range of instruments. However, as we have considered, the needs that various parts of the industry have for rheological information are limited by their ability to use the information. The typical situation one finds may be summarised as follows:

38

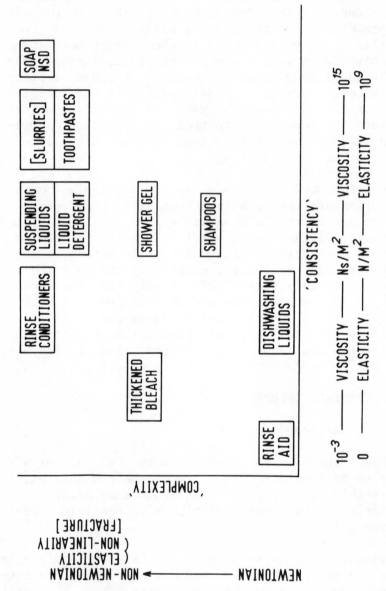

TABLE 1

	INSTRUMENT RESULT	CALCULATION
SHOP FLOOR	ONE READING	SIMPLE TABLE
QUALITY LAB.	FEW READINGS	SIMPLE TABLE
DEV. LAB.	COMPLETE CURVE	CALCULATOR
RES. LAB.	COMPLETE CHARACTERISATION	COMPUTER

TABLE 2

The following summary will cover the kinds of instruments covered in the table in terms of their levels of sophistication.

2.2 Summary of rheometers currently used

We may define a rheometer as an instrument that can:

 i) apply a deformation mode to a material and measure the subsequent force generated, or
ii) apply a force mode to a material and measure the subsequent deformation.

The best rheometer designs use geometries such that the forces/deformations can be reduced by subsequent calculation to stresses and strains and so produce material parameters. The sections that follow show how well various instruments meet these criteria.

2.2.1 Simple measurements

For many situations, the eye, hand (and sometimes the mouth) are sufficient to assess the rheological properties of detergent-like products - in fact these are the only facilities used by the consumer in product judgment! These inexact measurements produce not moduli but verbal descriptors of materials e.g. watery, thin, thick, stringy, jelly-like and so on. These are important in the area of consumer judgment,

RHEOMETER TYPE	VARIABLES TO CHANGE	MEASUREMENT	CONVENIENCE	ROBUSTNESS
	CAPILLARY DIAMETER	FLOW TIME fn (q)	*	*
	BALL SIZE	VELOCITY fn (q,k)	**	**
	EXIT TUBE DIAMETER	EFFLUX TIME fn (q,k)	****	****
	BOB SPEED DIAMETER	COUPLE fn (τ)	**	*

TABLE 3

but are very limited in producing an adequate technical description. However, for many years even major
detergent manufacturers relied solely on sensory descriptors of their products and raw materials.

When the need for quantifiable measurements was
recognised, industries naturally started with the
simplest, and the earlier applications were generally
in the quality-control area. Among the simpler types
of viscometers that were (and in many cases, still are)
used to produce a rudimentary rheological description
are i) the U-tube viscometer, ii) the rolling ball or
Hoeppler viscometer, iii) the Ford cup and iv) the
simplest "dip-in" rotary viscometers, e.g. the
original Brookfield viscometers.

These are used in an empirical manner, the choice
of variables being governed by ease of measurement;
i.e. if they are used in conjunction with a stopwatch
then a convenient timing would be in the range 5-50
secs, or if a scale displays a reading the variables
are chosen such that scale readings between 10 and 90%
of full scale deflection would be observed. Thus
convenience and not correctness of measurement were
the criteria of measurement. This is sufficient for
Newtonian liquids, but few detergent systems are
Newtonian.

Table 3 shows the variables available for changing
flow times and dial readings, together with some
indication of the type of measurement actually made,
the convenience of use and finally the instrument
robustness.

In using these "simple" measurements there is a
paradox in that the actual stress patterns and flow
fields are in fact very complicated. For instance,
in the U-tube viscometer, the stress in the liquid is
a function of geometry, position in the tube, time
and of course of the actual rheological properties.
Likewise, the rolling ball viscometer contains a complicated flow field with shear and extensional flow,
with the stress actually reversing in the liquid between the ball and the wall. The Ford cup viscometer

has a varying flow field with shear and extensional components and also with inertial forces often dominating. Another difficulty with the Ford cup type measurement is that the draining time of the liquid (which is used to calculate the "viscosity") is a function of fluid elasticity; often the fluid thread does not break into droplets, but produces a continuous thread until the cup is completely empty.

The greatest drawback of these instruments is however their inability to produce more than one number to characterise a fluid, even the Brookfield-type viscometer is often used at a fixed speed. This one number is usually converted into a "viscosity" via a Newtonian calibration. This can produce ambiguities as many different rheologies via shear and elongational viscosities and elastic effects produce the same apparent "viscosity".

A further drawback of these methods is that they do not produce compatible data, and thus, for instance, comparisons between different plants/factories are often impossible. Fig. 2 illustrates this point; Reng and Skrypzak show four typical products using three viscometers, only one of which had a known shear rate [1].#

In the next section simple modifications are suggested to update some of these methods to produce better data.

2.2.2 Modified simple methods

(i) The modified Hoeppler viscometer

The commercial instrument operates using a fixed tube angle and a variety of balls. The simplest modification is to use a variable angle of rolling. Altering the angle alters the average stress in the liquid produced by the component of the gravitational force on the ball in the tube direction.

Note: 1 cp $\equiv$ 1 mNsm^{-2} $\equiv$ 1 mPa.s

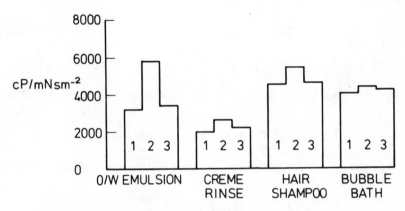

Fig.2. Comparison of different viscometers:

      1 Rolling ball viscometer (Höppler, ball 5)
      2 Rotary viscometer (Brookfield, spindle 2, 20 RPM)
      3 Rotary viscometer (Haake RV 3, $\gamma = 10$ sec^{-1})

(Taken from ref.1)

Sestak and Ambros [2] produced a one-dimensional pressure flow analysis of the flow through the constrictiction formed between the ball and the tube. For non-Newtonian Liquids that obey the power-law relationship:

$$\tau = \kappa q^n \, , \tag{1}$$

they produced a complicated relationship between κ, n and flow times for various geometries. These can be reduced to simpler forms such that:

$$\eta = K_1 \, (\sin \beta) T(A + Bn) \, , \tag{2}$$

where $A + B = 1$ and $\bar{q} = \dfrac{K_2}{T}$, β being the angle of the tube to the horizontal, n is the slope of the $\dfrac{\log \sin \beta}{\log \dfrac{1}{T}}$ curve, η is the viscosity at a mean shear rate $\bar{q}$ and T is the time for the ball to roll over a fixed distance. A, B, K_1 and K_2 are constants.

For a steel ball (diam. 0.907 cm) rolling in a precision glass tube (ID = 1.00 cm), with the ball timed over 10 cms for liquids of densities near unity:

$$\eta = 0.22 \, (\sin \beta) T(0.65 + 0.35 \, n) \, Nsm^{-2} \tag{3}$$

and

$$\bar{q} = \frac{10^4}{T} \, sec^{-1} \, . \tag{4}$$

These equations have been used to study various liquids (Newtonian and non-Newtonian) and the viscosity/shear rate curves are compared with those from a Haake Rotovisko concentric cylinder viscometer in Figs. 3 and 4.

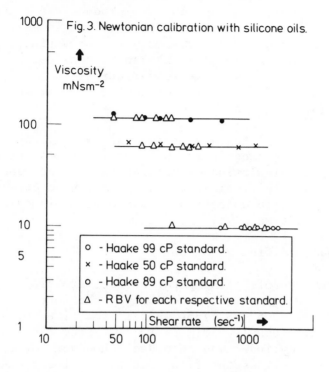

Fig.3. Newtonian calibration with silicone oils.

Viscosity
mNsm⁻²

o - Haake 99 cP standard.
× - Haake 50 cP standard.
o - Haake 89 cP standard.
△ - R B V for each respective standard.

Shear rate (sec⁻¹)

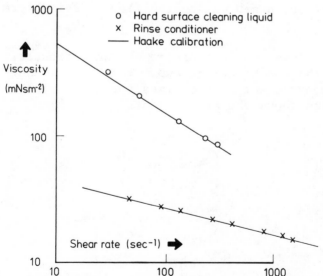

o   Hard surface cleaning liquid
×   Rinse conditioner
——   Haake calibration

Viscosity
(mNsm⁻²)

Shear rate (sec⁻¹)

Fig.4. Comparison of the rolling ball viscometer data & the Haake calibration
for two typical non-Newtonian detergent products
(using geometric values given in text)

This simple modification can be used with a minimum of two angles (say 30 and 60°) to produce the κ and n needed to describe the power-law liquid. In most cases, for a given product, κ shows much larger variations than n, so the factor (A + Bn) can be considered constant, because, as A + B = 1, variations in n (say ± 0.1) will produce little change for typical values of A and B, i.e. < 5%.

Quality control tests performed at two angles will produce all the description necessary for most products by a simple calculation extrapolating the two measured viscosities to unit shear rate. This can be done by means of a simple operation on a programmable calculator or using some simple nomogram.

(ii) The modified Ford cup

The drawback of defining the correct draining times of the cup was partially overcome by the introduction of the Ford A cup (BS 1733, 1955); the time for a fixed volume (usually 50 mls) to flow from the cup into a graduated cylinder was recorded. However for detergent-type solutions the flow into the cylinder causes foam formation, again making the flow time indeterminate.

These problems can be overcome by the introduction of pins into the inside of the cup. This allows the accurate measurement of the times involved. The following sketch shows the arrangement.

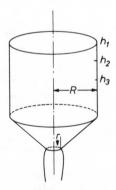

The Modified Ford Cup.

The original Ford B4 cup can be modified such that the original outlet (a small length of tube) is removed, and a simple sharp-edged orifice introduced. Two timing pins can then be introduced at heights h_2 and h_3 above the orifice.

Equating potential, kinetic energies and viscous dissipation, the following approximate flow times can be predicted for Newtonian liquids:

$$T_{ij} = \frac{3\eta}{z} \cdot \frac{R^2}{r^3} \cdot \frac{\Delta_{ij}}{\rho g} + \sqrt{\left(\frac{3\eta}{z} \cdot \frac{R^2}{r^3} \cdot \frac{\Delta_{ij}}{\rho g}\right)^2 + \frac{Bh_{ij}}{g}\left(\frac{R}{r}\right)^4 \Delta_{ij}^2} \;,$$

$$\ldots \quad (5)$$

where i, j refer to heights h_1, h_2, h_3 ,

$$\Delta_{ij} = \ln\left(\frac{h_i}{h_j}\right) , \qquad\qquad (6)$$

η is the Newtonian viscosity,
ρ is the fluid density,
R and r the radii of the cup and orifice respectively,
g the acceleration due to gravity, and

$$\bar{h}_{ij} = \frac{\sqrt{h_i} + \sqrt{h_j}}{2} . \qquad\qquad (7)$$

z and B are constants of order unity defining the relative contributions of viscous dissipation and kinetic energy production. They can either be evaluated by separate experiments or by fitting experimental data for very high and very low viscosity experiments.

Fig. 5 shows values of experimental flow time for a modified Ford cup type B4 (BS 1733), compared with those derived from the above equation, over the range 1 to ~ 750 mNsm^{-2}.

48

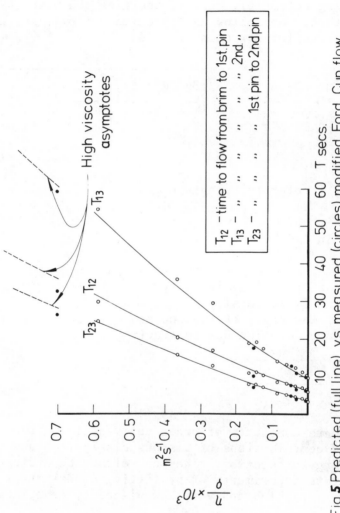

Fig. **5** Predicted (full line) vs. measured (circles) modified Ford Cup flow times for Newtonian liquids. Open and closed circles represent repeat measurements using different operator, stopwatch & raw materials.

For non-Newtonian (e.g. power-law) liquids, the viscosity will be:

$$\eta = \eta_{NEWTONIAN} \times \left(\frac{1 + n}{2}\right) \; Nsm^{-2} \tag{8}$$

and the average (equivalent) shear rate:

$$\bar{q}_{ij} = \frac{2(h_i - h_j) \; R^2}{T_{ij} \quad r^3} \; sec^{-1} \; . \tag{9}$$

Results obtained for a liquid scouring product are shown in Fig. 6 and for a heavy duty liquid detergent in Fig. 7, compared in both cases with concentric cylinder viscometer data.

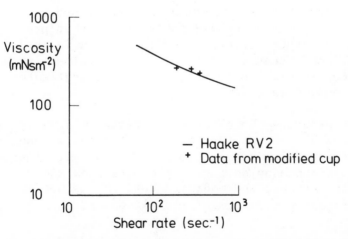

Fig.6. Comparison of data obtained from a modified Ford Cup with that from a rotary viscometer for a liquid scouring product.

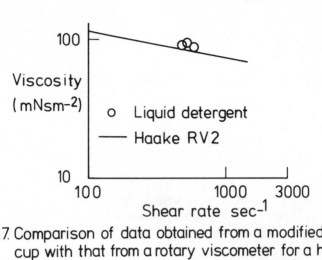

Fig.7. Comparison of data obtained from a modified Ford cup with that from a rotary viscometer for a heavy duty liquid detergent.

Products which show marked thixotropy produce flow times which can be considerably longer than predicted if steady state concentric cylinder data is used for comparison, because the total shear involved in the Ford cup is small compared with the steady state rotational data.

Also products where the elongational viscosity is considerably higher than the shear value (at equivalent deformation rates) can show anomalously long flow times. Since many detergent products show anomalous elastic effects, the modified Ford cup is not suitable for viscosity measurements in all cases.

(iii) The 'dip-in' viscometer - variable speed

The only defined shear-rate in the rotating bob system (with no outer container) is at the bob surface.

The shear-rate then drops off in a manner dependent on the rheological properties of the fluid.

For a limited range of flow models (e.g. power-law), various theories have been recently derived to produce κ and n from experimental data. Wein [3], for instance, has shown that for a rotating disc or sphere in a "power-law" liquid,

$$\tau = \frac{3M}{4\pi R^3} \; Nm^{-2} \text{ for both,}$$

$$q = \frac{8}{\pi n} \cdot \Omega \; sec^{-1} \text{ for a disc} \quad \text{and}$$

$$q = \frac{8}{\pi n} \cdot \left(\frac{3\pi}{4}\right)^{1/n} \cdot \Omega \; sec^{-1} \text{ for a sphere,}$$

where M is the couple, R the radius, Ω the rotation rate and n the power-law index which is equal to the slope of the $\log M / \log \Omega$ curve.

Other work [4] has also shown the usefulness of the Brookfield viscometer using a cylindrical bob.

2.2.3 Completely defined viscometers/rheometers

The shear viscosity of liquids can only be measured at defined shear rates using either cone and plate instruments with angles < 4° or concentric cylinder viscometers with a very narrow gap between the cylinders, otherwise in both cases there is a spatial variation of shear rate.

Among the commercial instruments available for such measurements (in the normal $1 - 1000 \; sec^{-1}$ shear-rate range) are the Haake Rotovisko, the Ferranti-Shirley viscometers and the Contraves Rotary viscometers. These instruments are widely used in the detergents industry. The typical advertising slogan "viscosity is a curve not a point, and you need a plot not a dot" has been appreciated by many. The use of such instruments in the detergent industry has been reviewed by Hoffman of Hoechst [5].

Other instruments which <u>can</u> operate in the simple
viscosity-shear rate mode are the "Deer" rheometer and
the Weissenberg Rheogoniometer, the former in a constant stress arrangement and the latter in a constant
shear-rate arrangement.

2.2.4 Other viscometers that allow calculation of average viscosity/average shear rate parameters

Capillary viscometers in which liquid is forced
through straight circular tubes are particularly useful for very high shear rates (up to 10^5 sec^{-1}). The
Mooney-Rabinowitz correction [6] allows (at least for
power-law behaviour) a definition of average viscosity/
average shear rate parameters. The data are useful in
assessing the behaviour of materials in high speed
mixers, spring loaded valves, etc. found in many
industrial situations.

2.2.5 Commercially available rheometers for rheological characterisation other than shear viscosity

A. The Deer Rheometer

The Deer Rheometer is a linear descendant of the
Stormer Viscometer [7] in that a stress is applied to
a material and the subsequent deformation is measured.
The Deer Rheometer allows a variety of stress patterns
to be applied (via a frictionless air bearing) e.g.

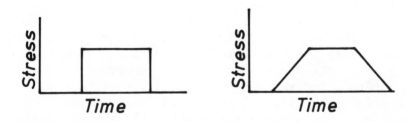

with variable time of application and rate of application of stress. Oscillatory stress application is
now available on the Petronics instrument.

The deformation can be monitored to measure:

(a) creep (via a chart recorder),
(b) steady-state shear-rate (via a meter),
(c) recoverable shear, and
(d) oscillatory parameters.

Various geometries are available to cover a range of applied stresses.

B. The Weissenberg Rheogoniometer

A complete Weissenberg Rheogoniometer can be used to measure:

(a) G' and G" as a function of frequency and deformation,
(b) η for a range of shear rates,
(c) ν_1 and ν_2 for a range of shear rates,
(d) τ and ν_1 as a function of shear/time at initiation or cessation of shear,
(e) G' and G" for a range of superimposed values of steady shear rate.

2.2.6 Orifice viscometers

Various instruments are available in the plastics industry to measure the parameters involved in pumping materials through circular sharp-edged orifices. These have been adapted by the detergent industries to deal with very viscous semi-solid materials such as soap. Cogswell [8] suggests that this flow can be approximated by a simple sink flow, enabling extension parameters to be assigned to the material.

2.2.7 Penetrometers

The penetrometer monitors the movement of a weighted penetrator into the surface of very viscous materials, e.g. bitumen, butter, etc. Typical indentors for use in soap are those developed for bitumen, being sharp truncated pins. The flow field around the penetrating pin is not yet well defined, but this instrument has proved very useful for soap-like materials.

2.2.8 On-line monitoring of viscosity

(i) Using pressure gradient

The pressure gradients produced by fluid flowing through successive pipes of differing diameters can be used to measure non-Newtonian parameters, if proper allowance is made for inlet and outlet effects.

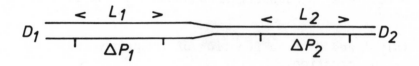

For a power-law fluid, the two pressure drops are given by:

$$\Delta P_1 = \frac{4 L_1 \kappa}{D_1} \left[\frac{8(3n + 1)Q}{\pi n D_1{}^3} \right]^n , \qquad (10)$$

$$\Delta P_2 = \frac{4 L_2 \kappa}{D_2} \left[\frac{8(3n + 1)Q}{\pi n D_2{}^3} \right]^n , \qquad (11)$$

where Q is the rate of flow. All the geometric parameters are known, the ΔP's and Q are measured by suitable means and thus κ and n can be calculated.

(ii) Using conventional viscometers placed in a
        stream of product

Various commercial models are available (e.g. Haake Viscontrol VA). They are usually concentric-cylinder type machines placed in the product line or in a by-pass line; slits in the outer cup allow product to be

continuously sampled in the shear gap. It is stated
in the trade literature that such a viscometer can
measure viscosities in the range $10 - 10^6$ mNsm^{-2}
depending on the shear rate. The drive system for
such instruments is usually composed of a normal
viscometer head coupled to a remote rotating member
via a non-contacting magnetic coupling. This allows
the rotating member to operate at extreme pressure
and/or temperature and in corrosive environments
without affecting the torque monitoring device. More
than one shear rate can be used. However it is doubt-
ful that the shear rate can be accurately defined with
product flowing across the shear gap continuously
(unless the flow rate is very slow) when various effects such as secondary flows and thixotropy can produce spurious results. Intermittent operation where
the flow is stopped (in a by-pass arrangement) to
take readings is the more effective method of use.

The Haake-type in-line viscometer has a facility for
controlling the mixing operation via "control contacts" which can respond when the viscosity falls
below or rises above pre-set maximum and minimum
values. This can be used in principle to control
viscosity by controlling the addition of a viscosity
modifying component (often a salt solution).

3. EXAMPLES OF THE USE OF RHEOMETRY

3.1 Introduction

In this section we review the use made of rheometry
in the detergents and allied industries and also those
academic studies of systems which are of interest to
the industries. The systems will be dealt with more
or less in terms of their consistency, going from
mobile "watery" systems through to soft solids. The
relevance of the rheological information obtained is
pointed out in terms of its contribution to basic
understanding, product formulation, consumer preference and chemical engineering calculation.

3.2 "Newtonian" systems

3.2.1 Very dilute surfactant solutions - basic studies

Anaeker [9] has pointed out that 'although small in number, viscosity studies involving surfactants have contributed valuable information about micelle shape'. The rheological measurement involves the accurate determination of viscosities just above that of water and usually long flow-time U-tube viscometers are used and it is generally assumed that the liquids are Newtonian.

Several theories are available to assess the shape of particles in solution, all depending on deviations in the viscosity from the simple Einstein equation. Oblate and prolate spheroid formulae are available which lead in the limit to discs and rods. Care has to be taken, however, in interpreting results because of possible charge and solvation effects. However, Stiger's model of the flexible rod-like micelle has been validated from viscosity data for dodecylammonium chloride. In principle this model can be checked using Kirkwood and Auer's expression [10] for the rigidity (G') at very high frequencies:

$$G' = \frac{6RT}{1000M} \ Nm^{-2} \ , \tag{12}$$

where M is the equivalent 'molecular' weight of the rod.

Gravshalt's [11] ultrasonic absorption and phase velocity measurements on highly dilute detergent systems is attempting this kind of approach. Her solutions contained molecular aggregates that differ from ordinary spherical and rod-like micelles. Pilpel [12] has used the G' expression to calculate the micellar molecular weights of potassium oleate in the presence of KCl and KOH.

Dilute solution viscometry has been used extensively to examine the molecular structure of polyoxyethylene, a typical nonionic surfactant [13]. At the very low molecular weight end of this series, the molecular

configuration of the polymer chains in solution can
be quite different from that of the high molecular
weight species, where the molecules are of sufficient
size that random flight statistics may apply. This
is found to affect the intrinsic viscosity. A clear
difference in behaviour is seen in the intrinsic
viscosity/molecular weight curve above and below
$Mw = 3 \times 10^3$. This has been observed for a number of
nonionic surfactant molecules (see ref. [13] p. 803).

The viscosity also reflects the great effect elec-
trolytes have on nonionic detergent molecules in
collapsing the polymer coil. The effect can be
quantified using the effective radius of gyration of
the polymer molecules on intrinsic viscosity. For
instance, polyoxyethylene ($Mw = 10^6$) has a calculated
RMS end-to-end distance of 1300–1800Å (depending on
the model used) in water at 35°C but only 620–770Å
in 0.45 M K_2SO_4 at 35°C (see ref. [13] p. 812).
These parameters can affect the rate of transport of
surfactant to the surface to be cleaned (e.g. clothes
in a washing machine), thus viscometric data is a
simple but effective means of assessing the molecular
parameters in the wide range of electrolyte conditions
found in wash liquors.

3.2.2 Detergent raw materials

Many detergent raw materials are themselves liquid
and Newtonian, for instance the "Tergitol" range of
nonionic and anionic biodegradable surfactants
developed by Union Carbide. The following table shows
the viscosity/temperature properties of the series
[14].

| Temp. | Viscosity ($mNsm^{-2}$) | | | | | |
| | Nonionic | | | | Anionic | |
	15-S-3	15-S-7	15-S-9	15-S-12	15-S-3A	15-S-3S
20°C	37	67	86	120	43	49
40°C	14	27	36	48	22	23
100°C	–	14	–	24	–	–

The structural formulae are:

NONIONIC $CH_3 - (CH_2)_n - CH_3$

$O - (CH_2 - CH_2O)_x$ Code is n-S-x

ANIONIC $CH_3 - (CH_2)_n - CH_3$

$O - (CH_2 - CH_2O)_x - SO_3 - NH_4$

If n=15 and x=3
then the code
is 15-S-3A

The viscosity increases with molecular weight (reflected in x) such that, for the anionic series, the viscosity is almost directly proportional to x.

In this series the "pour point" is given in $^{\circ}C$. Here we have an example of an ill-defined rheological parameter.

15-S-3	$- 30^{\circ}C$
15-S-7	$5^{\circ}C$
15-S-9	$10^{\circ}C$
15-S-12	$17^{\circ}C$
15-S-3A	$- 36^{\circ}C$
15-S-3S	$- 45^{\circ}C$

These figures relate to the temperature below which the product is not easily handled.

Recently Smith [15] of Continental Oil found a satisfactory theoretical relationship between room temperature viscosity of neat nonionic surfactants and the pour point, based on Eyring's kinetic approach; "an obvious practical feature of this work is that instead of the tedious relatively inexact practice of measuring pour points, it is apparent that one can obtain the same information from the more accurate and convenient measurement of viscosity". The model also gives an insight into the mechanisms which give rise to the pour point, and the molecular structure of neat individual nonionics and mixtures.

Addition of small amounts of water leads to large increases in viscosity of the 15-S-7, 9 and 12 nonionic Tergitols (which then become non-Newtonian) showing that for detergent type systems, viscosity is no guide to product "strength".

3.2.3 Rinse aids

These products are used to produce a 'shiny' appearance on crockery washed in automatic dishwashers; the consumer fills the automatic dispenser with product every few months. Their low viscosity aids good dispersion in the final rinsing water, and as the consumer is not actually dispensing the product, the low viscosity is not seen as a disadvantage. Typical product viscosities are in the range 15-30 $mNsm^{-2}$.

3.2.4 Shampoos and dishwashing liquids*

Shampoos offer a good illustration that viscosity is not an unambiguous guide to product strength. For instance, Barker et al. [16] give typical formulation details for a shampoo using a "high performance" surfactant. The viscosity of the product is greatly influenced by electrolyte concentration, even to a greater extent than the effect of the surfactant level. The 1% product with 3% salt is more viscous than the 2% surfactant product with 1% salt.

Rheometrical details in published formulation reports are rarely accompanied by actual details of procedure. Thus formulation details such as "This formulation has a typical viscosity of approximately 350 $mNsm^{-2}$ which can be increased to approximately 2000 $mNsm^{-2}$ by the addition of 2% sodium chloride" are rarely able to be duplicated fully in another laboratory using a different viscometer. In fact in the same article, the following instructions are given for

* Bubble bath products usually follow the same principles.

a similar formulation ... "Sodium chloride can be added to increase the viscosity, while if required, hexylene glycol may be used to decrease the viscosity" [17].

Dishwashing liquids are usually Newtonian liquids with active detergent concentrations around 40-50% [18]. The viscosities are in the range ∿100-250 $mNsm^{-2}$. Almost any viscometer may be used to characterise them, for instance the B4 Ford cup (BS 1733) would cover this range giving flow times between 35 and 75 secs.

Viscosity changes in this range are not easily noticed by eye but low viscosity products are often disliked by consumers who automatically (but often wrongly) infer that the product is dilute. The product viscosity is actually a complex function of active and electrolyte levels and chemical analysis is needed to find the product "strength".

Sherman [19] has reviewed consumer perception of viscosity and for those types of liquids which are assessed by pouring or shaking there is often a logarithmic relationship between subjective and instrumentally-measured viscosity such that changes in viscosity in the 100-200 $mNsm^{-2}$ range are difficult to notice.

3.2.5 Thickened bleaches

The ability to produce a thick rather than watery bleach gives the advantage of increasing the contact time for products draining from the sides of toilets and drains and the cleaning effect (i.e. the visual removal of soil) is greatly enhanced. Currently thickened bleaches have viscosities around 20 $mNsm^{-2}$.

Care must be taken in measuring these systems because of the extreme corrosive effect on normal viscometer components. To overcome this problem one has either to use special materials of construction (i.e. titanium) or coat the vulnerable parts present in some inert material (i.e. epoxy resin). The latter solution involves re-measuring the viscometer

dimensions, particularly with the small gap viscometers used for such low viscosities, e.g. the Haake
NV range.

3.2.6 Drag reduction

It is well known that low concentrations of high
molecular weight polymers can cause a substantial reduction in frictional drag in turbulent flow.   From
the wealth of data that now exists for a wide variety
of polymer-solvent systems (cf. [20]), there seems to
be a general agreement that the viscoelastic properties of the polymer are responsible for drag reduction
but the precise way in which the polymer interacts
with the solvent is still the subject of discussion
(although it appears that elongational viscosity plays
a major role). Polymers are not unique in forming
viscoelastic solutions of course and drag reduction
has been observed with other viscoelastic systems such
as micellar soap solutions and dispersions of fibrous
particles [21-25].

Micellar soap systems have not received the same
amount of attention as polymer solutions. With soap
systems, viscoelasticity and drag reduction have only
been observed with relatively concentrated solutions
whose viscosity is significantly greater than that of
the solvent so that the drag in the laminar flow
regime is considerably higher than with polymer solutions whose viscosity is indistinguishable from that
of the solvent. Previously it had been shown that by
using mixtures of surface active agents, viscoelasticity can be obtained at concentrations well below 0.5%
[26].

A very useful property of surfactant micellar drag
reducers is their insensitivity to continual shear.
The rod-like micelles which produce the elasticity
are dynamic entities that are continually exchanging
molecules with the surrounding solution. This means
that even if they are temporarily disrupted in the
turbulent sub-layer, they reform almost immediately
in a more quiescent region.

Surfactant solutions of the correct composition can show the necessary elastic effect over a range of temperatures, and could prove useful as long term drag-reducers in closed recirculating systems. Recently the British Gas Corporation sponsored work to investigate drag reduction possibilities in central heating systems [28]. The nonionic surfactant Lubrol 17 A10 (manufactured by ICI) in the presence of electrolyte (K_2SO_4) was found effective at temperatures near its cloud point. One interesting point we may note is the effect of temperature on the viscosity of these particular surfactant drag reducers, since above a critical temperature many show an increase in viscosity with increase in temperature (see Fig. 8). The optimum drag reduction was seen at a surfactant concentration of 0.25% in 0.45M electrolyte. Under these conditions, the viscosity (measured in a capillary tube type instrument) was only slightly higher than that of water at the same temperature; even small increases in surfactant level above 0.25% caused large increases in viscosity and the drag reduction decreased.

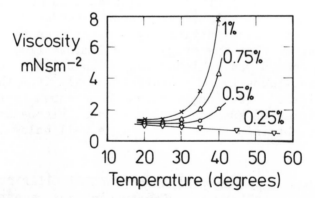

Fig. 8. Viscosity temperature data for Lubrol 17A10 +0.45N K_2SO_4 for various detergent concentrations.   (taken from ref. 28)

In summary, one might note that drag reduction in micellar systems only occurs when the micelles are rod-like - this will manifest itself also in enhanced viscosity and elasticity, thus suitable rheometry could screen potential systems using a minimum quantity of material. A full scale pipe flow rig operating at high temperature can be energy consuming and heavy on raw materials.

3.3 Medium viscosity non-Newtonian systems

3.3.1 Emulsions - basic studies

In emulsions, a critical role is played by the surfactant in emulsifying the disperse phase thus hindering coalescence. Sherman [29] has reviewed the history of emulsion rheology up to 1968 and its applications to industrial rheology up to 1970 [30]. The latest review of the general subject of industrial aspects of emulsion rheology appeared in "Cosmetics and Toiletries, July 1978 [31], where the usefulness of instruments such as the Weissenberg Rheogoniometer is emphasised, especially at very low shear rates and in oscillatory flow.

The work of Talman and co-workers to produce creams of various viscosity/shear rate and yield stress characteristics is noted, together with the later work of Barry using creep compliance. This type of experiment can now be performed on the commercially available Deer Rheometer. However, the usual (though not always appropriate) instrument used in this field is the rotational viscometer [32].

Komatsu and Takahashi [33] have examined W/O creams consisting of liquid paraffin and microcrystalline wax in the oil phase and polyoxyethylene-oleyl-ether as emulsifier. G' and G" were used to characterise the structure. They found that the elasticity G' arose from small water drops and the liquid paraffin, while the G" arose mainly from the large microcrystalline wax crystals in the liquid paraffin. The predominance of either G' or G" changes at a critical water concentration between 0.4 and 0.5 weight fraction.

Generally, in emulsions, the rheological parameters increase numerically with increasing phase volume, but sometimes above a critical phase volume inversion occurs and the viscosity and rigidity undergo a drastic reduction because the inverted phase volume is much smaller.

The structure of emulsions at very low shear rates can be examined by performing tests in the type of instrument suggested by Zimm and Crothers [34]. Shear rates down to 10^{-6} sec^{-1} are then possible, thus allowing accurate measurement of the true yield stress. The yield stress data from this measurement may be compared with various theoretical models where the van der Waals forces between the interacting particles have been used to calculate a yield stress [35-37].

Recently, Lyklema et al. [38] have used a similar interaction model, but with strong and weak bonds. This model seeks to interpret creep data on emulsion systems measured in a concentric cylinder instrument, where a constant couple was applied to the inner cylinder, while its deformation was measured by a displacement transducer [39]. Stresses of a few Nm^{-2} could be applied and strains down to about 0.01 measured. The weak bonds were assigned to van der Waals attraction between particles and the strong bonds to polyelectrolyte attraction between the emulsifier layers around the particles. The model yielded a linear relationship between strain and log time which is often seen for such systems.

The use of rheometry to produce a quantitative description of emulsions that can be compared with qualitative consumer response has been reviewed by Moskowitz and Fishken [40]. They list various verbal descriptors such as "viscosity, cohesiveness, springiness, gumminess, adhesiveness" and note that these quantities as perceived by panellists are relatively insensitive to actual rheometer values of η, G' etc. A ten-fold change in measured viscosity appears only as an approximate 2.5→3-fold change in perceived "thickness" or "thinness". They further note that, together with colour and fragrance, "viscosity" is a

key parameter in consumer assessment of creams and has
to be optimised to maximise consumer purchase interest,
see Table X in ref. 40 (see also Sherman's review in
SPC Jan. 1972, p.54).

Long term emulsion stability tests are very important to ensure the consumer actually obtains what the
manufacturer made. Various accelerated storage tests
have been used where the rheological properties are
examined after various accelerated ageing processes
such as exposure to elevated or depressed temperatures
and high speed or ultra centrifugation [41].

Roehl reports the use of a Brookfield (type unspeci-
fied) viscometer at rpm's 1.5 and 60 to assess the
effect of homogenisation on the viscosity and thixotropy of creams and from this concludes that subsequent long term measurements of this form can be used
for monitoring changes in particle interaction. He
also recommends that penetrometry be used for creams
of high consistency; the DIN 51804 standard, cone
weight 102.5g penetrating for 5 secs being particularly recommended.

Lastly we note the work of Eccleston using three
rheometers for concurrent studies on emulsion systems
[42,43]:

1. The Ferranti-Shirley cone-and-plate viscometer in
the automatic mode, sweeping the shear rate range 0-
1671 sec^{-1} while monitoring shear stress,
2. Concentric cylinder reaction air turbine creep
rheometer (the forerunner of the Deer Rheometer)
giving compliance vs time, and
3. The Weissenberg Rheogoniometer, used to measure
the amplitude ratio and phase lag (from which in principle G' and G" can be measured) as a function of
frequency from 10^{-2} to 10^{1} Hz, using the parallel
plate arrangement with a gap of 0.635 mm.

The creep rheometer and rheogoniometer data (J the
compliance as a function of time and G' as a function
of frequency) were shown to be interchangeable to give
either kind of data over a large time or frequency

range (i.e. 10^{-8} Hz to 10^{2} Hz).

On using the rheometers to study a mixed (surfactant/ fatty alcohol) emulsifier for oil in water emulsions, great similarity was noted between the systems with and without the dispersed oil phase. This led to the hypothesis that the rheology was controlled by a "gel" continuous phase into which the oil could be dispersed to form the emulsion. This gelled continuous phase was postulated to be due to a liquid crystalline type phase formed by the surfactant/fatty alcohol interacting with the water.

This model is not too dissimilar to that of Jansson and Friberg [44] who postulate that emulsions are stabilised by absorbed lamellar liquid crystals.

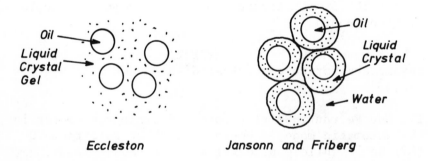

Eccleston   Jansonn and Friberg

3.3.2 Fabric softeners

Fabric softeners have the effect of reducing the static charge on man-made fabrics and give a softer feel to cotton articles [45]. They usually contain from 4-8% cationic detergent material and are pourable easily-dispersed liquids [46]. Buckering and May report that the manufacturing conditions have a great influence on the appearance of the finished product.

First a gel is prepared by heating the cationic with a part of the water to 50°C and the gel is then diluted with the rest of the water. The mixture is stirred throughout the operation and the mode of stirring (duration, speed, type of mixer) conditions the final viscosity of the softener. The viscosity should not change too much on storage. Small amounts of sodium chloride or acetate are used to cause a lowering of viscosity, while the addition of methyl cellulose or long chained alcohols increases the viscosity [47].

Recently James and Ogden [48] of Akzo Chemie, a major supplier of raw materials for this product, have shown that the microstructure that gives the high viscosity of these relatively low concentration products is due to the presence of multiwalled vesicles:

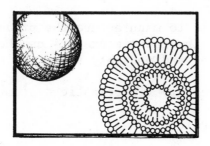

The relative amount of material and water in the disperse phase, and hence the phase volume and thus viscosity, depends on the particular conditions under which the dispersion is prepared. If small amounts of ionic or nonionic materials are present, slow osmotic swelling or shrinkage of the vesicles can occur leading to marked and often disastrous changes in viscosity on storage.

In their study, James and Ogden employed a Brookfield RVT viscometer using a spindle speed of 100 rpm in all measurements "in order to minimise variation due to shear thinning", showing that rheometrical problems are appreciated if not solved.

Little or no data are available in the open literature on actual viscosities, viscometers, shear rates,
etc. for the type of studies quoted above. However,
if one surveys currently available products, a wide
range of viscosity/shear rate behaviour is seen, see
Fig. 9A. It will be observed that more viscous products are more non-Newtonian and vice versa, showing
a typical Nodonchelle-Schutz type behaviour [49], (cf.
Fig. 9B, where the power law parameters κ and n are
measures of the viscosity level and "non-Newtonianness"
respectively.)

This product is particularly susceptible to storage
at low temperatures, and if no antifreeze additive is
used, the product can be "squeezed from the bottle
like toothpaste" [50] after freezing. This thickening
is irreversible. The over-addition of electrolyte in
manufacture for viscosity control can cause irreversible "salting out" of the product.

It is possible to manufacture the product continuously [51] and in-line viscometers can be used for
quality checking and possibly viscosity control.

3.4 Medium viscosity viscoelastic systems

3.4.1 Basic studies

Many of the detergent based elastic liquids favoured
by rheologists because of their very strong elastic
effects fall into this category [52].

Although the viscosity of soap/electrolyte systems
has been of interest for a long time [53], progress
in these systems started in earnest in the late 1940's
(e.g. Bungerberg et al. [54]). This was extended in
the 1950's by Winsor [55] and Pilpel [56]. In 1966
Pilpel [57] published a study of the rheology of elastic detergent systems using the Weissenberg Rheogoniometer. Previously, only visual recoil had been noted;
this study put the elasticity on a quantitative basis.
Normal and shear stresses at shear rates between 1 and
20 sec^{-1} were measured using $\frac{1}{4}^{\circ}$ and $\frac{1}{2}^{\circ}$ cone-and-plate
configurations with plate radii of 5 and 10 cms. For

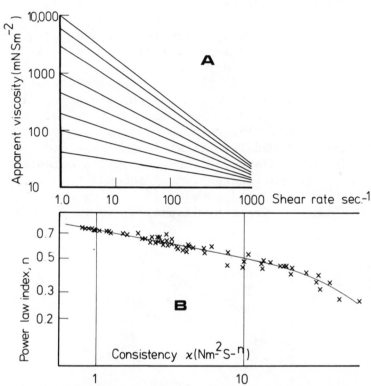

Fig. 9. A.) Typical viscosity - shear rate curves for rinse conditioners
B.) x, n data for 60 rinse conditioners

normal stresses < 10 Nm^{-2} the readings were checked on a capillary tube attachment, measuring the height of liquid in tubes mounted on the top plate. (This can lead to large errors due to "hole" effects – see Walters' "Rheometry" [144, p. 88] on the effect of hole errors on the first normal stress in a detergent based viscoelastic liquid.) Pilpel also measured the oscillatory behaviour using a resonance method.

Figs. 10 and 11 show the results obtained for a potassium oleate/potassium chloride mixture. Pilpel showed that this and other systems [57] produce their maximum in viscosity and elasticity at a particular electrolyte level just prior to a phase change; (in the 1966 work this was coacervation and/or "salting out" of the soap). The results from the study were used to test two models: first an emulsion model using Oldroyd's theory [58] and secondly a network model based on the interaction of secondary soap micelles based on a Lodge model [59] using thin rod-like micelles (or prolate ellipsoids) as the network

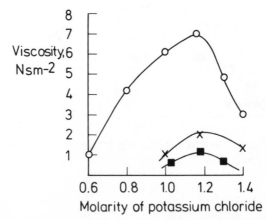

Fig.10. Viscosity – Composition data for Potassium chloride/
potassium oleate solutions, shear rate ~ 1sec.$^{-1}$

   o 0.035 M Potassium oleate
   x 0.02 "   "   "
   ■ 0.015 "   "   "

(Data taken from ref. 57.)

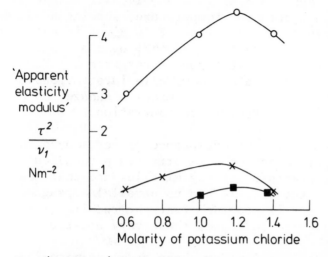

Fig.11. `Elasticity´- composition data for potassium chloride/ potassium oleate solutions, shear rate ~ 1sec.-1 (Symbols as fig.10.)

(Data taken from ref 57)

components. The latter model was found to be more consistent with the results. Using [10]

$$G' = \frac{3RTC_O}{5M} , \qquad (13)$$

where M is the molecular weight of the "rods", and C_O their concentration in g/cm^3, Pilpel found that the equation predicted values of M consistent with Debye's model [60] of cylindrical micelles. Micelle molecular weight in the high viscosity (optimum salt) region was calculated to be $\sim10^6$ thus giving a length of order $10^3 \overset{\circ}{A}$ and a diameter $\sim50\overset{\circ}{A}$.

Den Otter and Papenhuijzen [61] studied a 10% sodium lauryl poly(oxyethylene) sulphate solution in 2.4 and 1.7 molar $NaNO_3$ over a range of oscillatory frequencies and found a maximum in G" and a plateau in G' at high frequency. It was concluded that over the range of frequencies studied, the response of the system could be described by a single relaxation time Maxwell model! Estimates of the micellar molecular weight gave values between 10^6 and 1.8×10^6. The molecular weight decreased with increasing temperature, showing a decrease in the length of the micelle. Flow birefringence and viscosity curves were measured [62] over a range of $NaNO_3$ concentrations at temperatures from 5-34°C and the results were related to micelle interactions. At > 0.9M $NaNO_3$, the size of the micelle changed monotonically with electrolyte concentration.

The negative flow birefringence effect showed that the detergent molecules were oriented perpendicular to the largest dimension of the micelles and that the molecules in the micelle had no parallel arrangement. Thus more proof is provided that the rod-like (or cylindrical) micelle causes these very elastic phases.

Further confirmation for these rod-like micelles causing elasticity comes from the NMR work of Tiddy and Wheeler [63]. They considered the formation and breakdown kinetics of such entities on dilution from more concentrated phases, and were able to show from NMR and light scattering that the viscoelasticity was simply a reflection of the presence of rod-like micelles and that the breakdown of these into spherical micelles caused the viscoelasticity to disappear.

Saul et al. [64], again using NMR, showed that maxima in NMR signals could be correlated with those in viscosity and first normal stress difference measured on the Weissenberg Rheogoniometer. Similarly when using different concentrations of the same rod-like micellar solution, they were able to show that the first appearance of these micelles also corresponds to the first appearance of visual recoil (i.e. viscoelasticity), see Fig. 12.

Similar observations have been made by Jones [65] and Zettlemeyer [66]; again NMR data suggests that viscoelasticity is a result of the presence of rod-like micelles. Hyde [67] reported a similar study in which the rheological properties were measured over a range of temperature, concentration and shear rates.

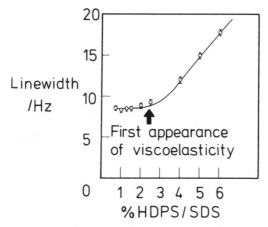

Fig.*12*.N.m.r. line width against surfactant concentration of HDPS/SDS Mixtures (ratio HDPS:SDS = 3.5: 1.5), showing position of first appearance of visible recoil (viscoelasticity).

Taken from ref. 64.

A complete rheological characterisation of the shear properties of two concentrated rod-like micellar solutions was carried out by Barnes et al. [68]. Using the Weissenberg Rheogoniometer in the steady and oscillatory modes, G', G'', η and ν_1 were measured as a function of frequency and shear rate. The systems were:

- 0.05M/0.05M solution of cetyltrimethylammonium bromide (CTAB) and toluene sulphonic acid (TSA) and
- 16% tetradecyl dimethylammonio propane (DMH) and 4% sodium dodecyl sulphate (SDS) (percentages by weight).

A further study [69] covered the measurement of the second normal stress difference ν_2.

Figs. 13 and 14 show typical data for these systems; for comparison, two polymer systems are also shown – one water based and the other oil based.

74

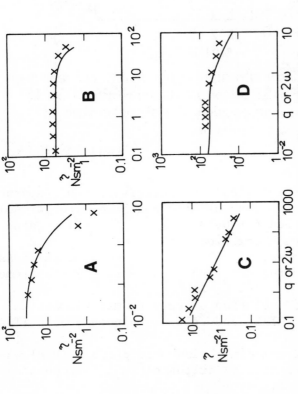

Fig.13. Viscosity vs. shear rate or frequency for detergent & polymer solutions using an R18 Rheogoniometer, solid line steady shear, x-oscill.

A CTAB/TSA detergent mixture.
B DMH/SDS   "   "
C 1% Polyacrylamide in 70/30 glycerine/water.
D Paratac-polyisobutylene in mineral oil.   (Taken from ref. 68)

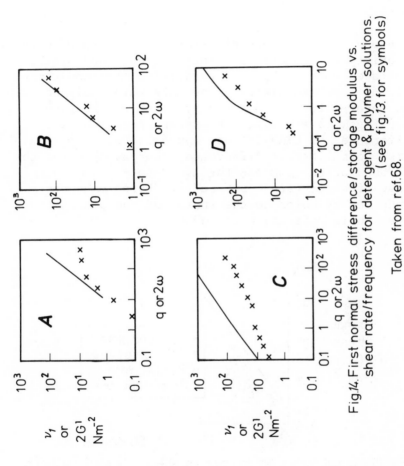

Fig.14. First normal stress difference/storage modulus vs. shear rate/frequency for detergent & polymer solutions. (see fig.13. for symbols)

Taken from ref.68.

The detergent solutions exhibited the same general rheological phenomena usually found in polymer solutions and could be interpreted in terms of models developed for polymer systems, especially those found successful and based on inter-molecular forces [70,71]. Again, the apparent molecular weight could be calculated (using these theories) from all the linear data. Values were found to be in the range 3×10^6 to 7×10^6.

Second normal stress differences were measured using a variety of geometries – touching cone and plate, separated cone and plate, parallel plates and re-entrant cone and plates. The data were analysed using the general formula derived by Marsh and Pearson [72]. (see Figs. 15 and 16). The second normal stress difference for CTAB/TSA was approximately $-\frac{1}{10}$ of the first normal stress difference, whereas that for the DMH/SDS varied from $+\frac{1}{4}$ through 0 to $-\frac{1}{3}$ over the range of shear rate 6–50 sec^{-1}*. Using manometers connected by holes to the inside of the plates, the usual "hole" effects were found.

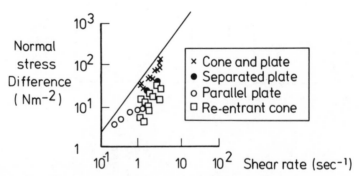

Fig.15. Normal stress differences vs. shear rate for CTAB/TSA. Solid line is the level of the first normal stress difference ($\times$-1). Symbols refer to geometry for second normal stress difference measurement:

* Typical values for polymer systems are approx. $-\frac{1}{10}$ [69].

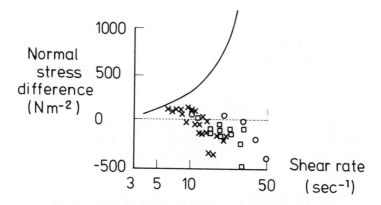

Fig.16. Normal stress differences vs. shear rate for
DMH/SDS. Solid line is first normal stress
difference: see fig.15 for symbols.

Taken from ref. 69.

Recently Doi and Edwards [73,74] published a theoretical treatment for interacting rods as a description
of possible macromolecular solutions. This could
predict all the rheological phenomena usually measured
experimentally. As the model predicts all the
phenomena from the physical hindrance of the rods to
each other, and with no specific interaction between
them, the theory is relatively easy to use for a wide
range of different rod-like systems.

More complicated rheological testing on a (probably inverted) rod-like micellar system was carried out by Boiij [75], who studied a 10% solution of aluminium di-laureate soap in toluene under combined steady and oscillatory shear and showed a typical polymer-like behaviour.

Some nonionic detergents are actually macromolecules with molecular weights of around 10^6, e.g. polyoxyethylene, and in such cases non-Newtonian/viscoelastic behaviour can arise without association into micelles. Data taken for such a system (MW = 5.5×10^6) using an Ubbelohde-type multibulb capillary viscometer are shown in Fig. 17. From this plot it is clear that shear dependence must be accounted for, even in dilute solutions. This behaviour has been associated with the existence of a 'temporary' network [76].

Higher concentrations of such systems exhibit 'stringiness' from about 0.5 to 5%. This polymer surfactant has been studied at these concentrations in a Bergen-type viscometer [77,78]. Fig. 18 shows the large dependence of viscosity on shear rate. An automatic recording rotational viscometer was used to measure these systems at high shear rates. Fig. 18 shows that shear thinning continues to above $4000 \, sec^{-1}$. Rodriguez and Gaettler [79] found that this obeyed the empirical viscosity/shear rate relationship over five decades of shear rate:

$$\eta/\eta_o = 0.68 - 0.32 \; erf \left[\frac{[\log \tau q - \log B]}{3.21} \right] . \qquad (14)$$

Erf is the error function, τq the energy dissipation, B the values of τq at the inflection point in the flow curve. B depended only on molecular weight and η_o on concentration and molecular weight.

Electrolyte has the opposite effect on polymer surfactant to that on most micellar systems, for here a 2% aqueous solution of polyoxyethylene dropped from $230 \, Nsm^{-2}$ to less than $10 \, Nsm^{-2}$ when in 1M KCl solution.

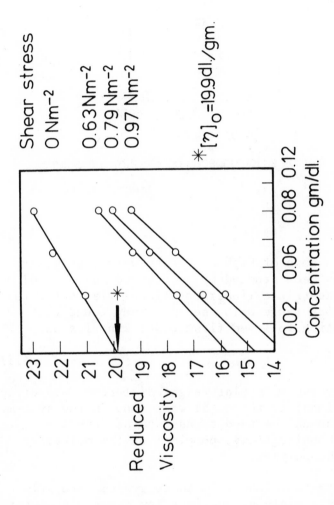

Shear stress

0 Nm⁻²

0.63 Nm⁻²
0.79 Nm⁻²
0.97 Nm⁻²

* $[\eta]_o = 19.9$ dl./gm.

Fig. 17. Reduced viscosity of polyoxyethylene at various shear stresses; 45°C, $M_w \sim 5 \times 10^6$

Taken from ref. 76.

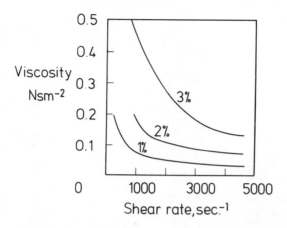

Fig. 18. Viscosity of shear, aqueous polyoxyethylene solutions.

Taken from ref. 76.

Shear breakdown of high molecular weight polyoxyethylene has been studied by Asbeck and Baxter [80]
using a concentric cylinder rotational viscometer in
the shear rate range $0-10^3$ sec^{-1}. They found that
the degradation followed first order kinetics and

$$M^{-1} - M_o^{-1} = K\eta q^2 t \ , \tag{15}$$

i.e. the inverse molecular weight difference was directly proportional to the total work done on the system.
This equation can be used to assess the effect of
mechanical handling (e.g. pumping) on the molecular
weight of the polymer.

At concentrations above 5% these systems are gels
that exhibit a yield stress in a low shear viscometer
[81].

3.4.2 Shower gels and gel-like hair shampoos

The recent boom in the use of showers has created a
product opportunity for manufacturers. A number of

shower gels, body shampoos and so on have now appeared
in the market place. These have the obvious prerequisite that they should be easily dispensed and distributed over the body, but should not be so "thin" that
they flow through the fingers. They share the same
rheological requirements as gel-like hair shampoos –
indeed many such products are sold as dual purpose,
body and hair shampoos.

The properties of detergent phases make it possible
to produce these (pseudo) gels using merely the detergent itself and small amounts of inorganic materials,
e.g. salt.

A true gel of course has a yield stress but these
"gels" have very high low-shear-rate viscosities and
high elasticity which give a jelly-like appearance.
However, unlike jelly, they need to be easily broken
down in shear but able to rebuild the structure
quickly.

Although quality control of such materials would
need to measure viscosity, elasticity, thixotropy,
etc., one finds that all these parameters (for any
given formulation) are usually "scaled" by the viscosity, so this alone is often a good enough measure of
product quality.

The rheology and indeed the practical application
of these 'gel' systems bear close resemblance to their
inverted oil-based analogues, i.e. greases (cf.
Chapter 3).

3.5 Moderately viscous systems with a yield stress

3.5.1 Liquid abrasive cleaners

Liquid abrasive products depend (for their technical
success) on their ability to suspend small abrasive
particles indefinitely. A typical abrasive particle
has a diameter of $\sim 50 \mu m$ and a density difference compared with the suspending phase of ~ 1.5, thus the
yield stress needed to suspend the particle is about
$0.5 \ Nm^{-2}$ [82]. This constraint, together with the need

to pour and dispense the product, gives the rheological specification for the product.

The measurement of the yield stress at very low shear rates presents considerable problems unless very sophisticated low shear equipment is available (e.g. the Weissenberg Rheogoniometer or the Deer Rheometer). Problems even arise with the Rheogoniometer because the abrasive particle size can be large compared with the gap formed between the (truncated) cone and plate.

Problems arise in trying to produce simple and rapid methods of measuring yield stress for quality control purposes by relatively unskilled personnel and so far no details have been published on such tests. Attempts have been made using the stress value obtained by extrapolating normal viscometer data to zero shear rate, but this has not met with great success [82].

Liquid detergents have a similar need for a yield stress to suspend the excess undissolved particles of "builder" used to reduce water hardness. However, the level of yield stress needed is much lower than that for liquid abrasives, i.e. ~ 0.05 Nm^{-2}. Such yield stresses are very difficult to measure even with very sophisticated equipment.

3.6 Viscous non-Newtonian systems

3.6.1 Basic studies - liquid crystals

In a personal communication, Tiddy [83] concludes that "at the present time there are no descriptions of liquid crystal rheology that relate observed viscosity behaviour to the properties of the fundamental structural units ...".

The rheology is obviously dominated by the structure of the phase, but also [84] by the defects in the structure, because real liquid crystal phases are made up of crystallites with random alignment rather than indefinitely long sheets of structured material. Alignment of the crystallites with shear has been proposed as a mechanism for shear thinning [85-90] at low shear rates. At higher shear rates, the creation of more defects can lead to the viscosity increasing with shear [91,92].

In measuring the rheology of liquid crystals, great care has to be taken to treat the rheometer surfaces to produce a consistent relationship between the liquid crystal molecules and the shear surfaces. This can have a large effect on the measurements made in small gap instruments. This crystal/surface inter-action controls the orientation of the 'director' vectors which gives the local state of molecular alignment. The boundary conditions are then fixed. As an example, Kahn [93] used a silane coupling agent (Dow Corning 2300) to produce homeotropic orientation of liquid crystal molecules on glass surfaces. This boundary dominated structure can lead to a situation where in capillary flow, for example, two regions of flow behaviour can occur; first a wall region where there is a preferred alignment imposed on the mole-cules due to the wall and secondly a core region where the flow is dominated by the interaction of the liquid crystal and the flow. So there can be a change in the orientation from a perpendicular orientation at the wall to a direction more or less parallel to the tube axis in the core region. This can produce a strong dependence of apparent viscosity on tube radius. Fisher and Fredrickson [94] have derived a simple expression to relate the apparent viscosities in the two regions:

$$\frac{R - \delta}{R} = \frac{\eta_{core}}{\eta_{APP}} \left(\frac{\eta_{wall} - \eta_{APP}}{\eta_{wall} - \eta_{core}} \right)^{\frac{1}{4}} , \qquad (16)$$

where δ is the thickness of the wall region.

White et al. [95] quote values of δ between 6 and 7%. As the shear rate and thus shear stress is largest at the wall, even a small boundary layer of diverse material can cause an appreciable difference in apparent viscosity.

The problems associated with interpreting flows
governed by wall effects are compounded by the effects
of random defects in the crystal structure, with the
extent and nature of the defects actually depending
on how the sample is placed in the rheometer. For
instance, Horn and Kleman [91] describe a modified
Weissenberg Rheogoniometer where the number of defects
can be studied in situ using a glass cone and plate
arrangement and a high quality optical microscope.
If the sample is carefully loaded with the layers
aligned parallel to the surfaces, the viscosity is
low, i.e. the layers slide over each other freely.
However, if the layers are buckled, shearing creates
defects in this structure, and prolonged shearing
produces a completely disordered arrangement of crystallites. This produces a much higher viscosity.
Fig. 19 shows the effect as seen in the shear stress/
shear rate curves. The original parallel arrangement
was produced by an annealing process produced by
persistent oscillation which removes all the defects.

Bohlin and Fontell [96] report similar time/shear
rate dependent viscosities using a Contraves Rheomat
30 for steady state experiment and a Contraves balance
rheometer for oscillatory testing. Using a CTAB-
hexan-1-ol-water liquid crystal system they observed
an actual reformation of structure, where shearing
caused a more ordered phase and increased the viscosity and elasticity parameters.

This change of phase with shearing phenomena has
also been reported by Gravsholt [97]. Using a single
component system (cetyl trimethyl ammonium salicylate)
at very low concentrations (0.2-0.3 mM), shearing at
low rates showed Newtonian behaviour. However, at
higher shear rates (> 2 sec^{-1}), rheopectic behaviour
was seen, see Fig. 20. Hyde et al. showed the same
behaviour but at higher concentrations ($\sim$30 mM) and
much higher shear rates ($\sim$ 1000 sec^{-1}) [98].

The picture then is one of great complexity, first
in actually interpreting rheometer data which can
show geometry and time effects and then relating
these effects to the microstructure. All these

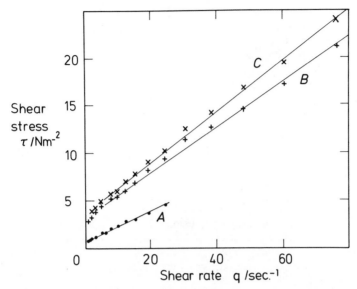

Fig.*19.* Shear stress/shear rate for a liquid crystal sample (n⁻octyl⁻
cyanobiphenyl) at various defect levels.

A Zero defects
B 75% defects
C 90% defects

Taken from ref. 91.

effects can be seen in real detergent systems and
cause difficulties in formulating and processing
products.

Soap liquid crystal phases are known as "smectic".
Single component soap systems possess a high viscosity
and considerable rigidity [99-101]. Glasstone suggests
that normal liquid flow does not occur, but movement
is caused by the gliding of planes over each other
[102].

Extrusion plastometer measurements at 140°C on a
series of pure straight-chain sodium soaps of varying

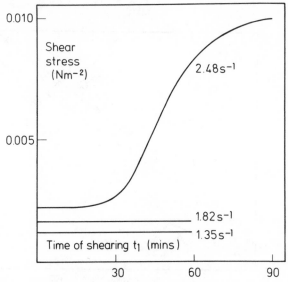

Fig. 20. Shear stress/time curves at various shear rates for 0.45mN CTA-Sal solution at 25°C.

Taken from ref. 97.

molecular weight show an abrupt increase in the yield stress when the number of carbon atoms in the soap is below eight [103]. Puddington, using a similar technique, showed that anhydrous soap phase changes could be followed by changes in the viscosity. Large changes in soap mobility were seen at 68 and 105°C. The viscosity becomes Newtonian above 298°C [104]. Other aspects of soap and liquid crystal rheometry are well covered by Porter and Johnson in their contribution to "Rheology" [105].

In industrial situations, the phase behaviour and thus the rheology of soap and detergent phases is complicated by the broad molecular weight distribution of the materials, considerable amounts of impurities such as inorganic salts being present. Also, in products, they are mixed with other organic and inorganic materials that greatly affect their behaviour.

3.6.2 Detergent slurries

Detergent powders are manufactured by spray drying highly concentrated slurries of the detergent components dispersed in excess water. These are usually manufactured and handled at about $80^{\circ}C$. To make the spray drying process as economical as possible, the maximum amount of solid material has to be dispersed in the slurry so that a minimum amount of water needs to be evaporated. This of course is balanced by the need to keep the slurry at a viscosity low enough to mix and pump.

High active matter detergent slurries (containing up to 35-45% phase volume of solids) are extremely viscous and often attempts are made to find "magic" ingredients to reduce the viscosities. The addition of a "hydrotope" such as sodium xylene sulphate often assists in reducing slurry viscosity [106]. This probably destroys the ordered structures often encountered in detergent systems. The effect of xylene sulphonate on a 45% detergent slurry at $60^{\circ}C$ (containing 20% sodium alkylbenzene sulphonate; 40% sodium pyrophosphate and 40% sodium sulphate) is shown in Fig. 21. These additives also help improve the storage properties of the final product by acting as an anti-caking agent, reducing the tendency for the formation of aggregates. This also makes the final powder more dispersible, avoiding the deposition of undispersed powder on the washed articles [107].

Typical viscosity/shear rate curves of handleable slurries are shown in Fig. 22. Generally we could classify slurries as "thick, average, and thin" according to their ease of handling using the following criteria:

	$K(Ns^{n}m^{-2})$	n (-)
Thick	>100	0.2
Average	1-10	0.5
Thin	<1	0.7

Only 'average and below' slurries can be atomised under normal conditions.

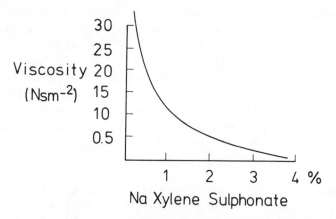

Viscosity (Nsm^{-2})

Na Xylene Sulphonate

Fig.21. Change in the viscosity of a detergent slurry
induced by the addition of a sodium xylene
sulphonate hydrotrope.

Taken from ref. 106

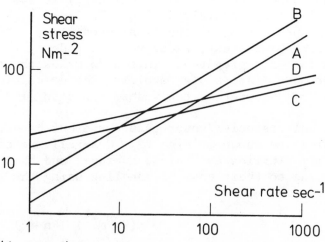

Fig.22.(Averaged)shear-stress/shear-rate curves for four
typical detergent slurries at 80°C. measured in a
Haake viscometer (Rotovisco-S V arrangement).

The overall power requirements for mixing slurries can be calculated (cf. [108-111]). It appears that non-Newtonian liquids are more difficult to mix thoroughly than are the equivalent Newtonian fluids.

Heat transfer in flowing non-Newtonian fluids has been reviewed by Edwards and Wilkinson [112] and Astarita and Mashelkar [113] and others [114], and here again similar treatments are available using κ and n, obtained from viscometric data. These calculations are important because detergent processing is carried out at high temperatures.

3.6.3 Toothpastes

The change in tooth-cleaning products from powder to semi-fluid gave an obvious improvement in their application, first to the toothbrush and secondly to the teeth themselves.

The ideal toothpaste should have a yield stress when unsheared, to avoid slumping when squeezed onto the toothbrush, but must be shear thinning to provide easy removal from the tube.

Block [115] has summarised the usual empirical tests used for assessing the yield stress and extrusion properties of toothpastes. First, he describes a "sag-meter" to measure the slumping effect; he notes "... it is a common experience that after extrusion from a tube onto a toothbrush, some toothpastes tend to "sag" or slump between the bristles. With pastes of low dispersibility, sag would increase the time required to clean the toothbrush after use. The slow ooze of toothpaste from an open tube left in the bathroom is probably also associated with the property of sag. From these considerations it seems likely that the overall preference for a toothpaste may suffer if the property of sag falls below a certain limit." Block then proceeds to describe an empirical measurement of "sagginess", where toothpaste strips of various thickness are allowed to sag under their own weight. A sag number is assigned to the paste depending on the minimum thickness necessary for a toothpaste to sag visibly.

A time dependence to reach a certain ultimate sag
number is also noted, thus both modulus and viscosity
are empirically monitored at different stresses.

These measurements can be carried out on instruments
where the deformation field is known, so that actual
moduli and viscosities can be assigned. The Deer
Rheometer using variable stress or the Weissenberg
Rheogoniometer could be used, but of course at much
greater cost. The empirical measurement always has
the attraction of low cost and usually easy operation!

To measure the extrusion properties, Block describes
a simple test performed on an Instron tensile tester
fitted with a plunger to extrude toothpaste through a
simulated toothpaste tube nozzle. If care is taken,
an actual tube of toothpaste can be used after cutting
off its crimped end and inserting the plunger.

Tensile strength can be measured for toothpastes by
extruding vertically downwards. At a critical length
the extrudate breaks off, thus giving a value for the
strength. The strength increases with storage time
for most samples, presumably due to re-structuring
after the toothpaste encountered being mixed,
pumped and filled. This strength is probably close
to the yield stress in extensional flow. Fig. 23
shows the stress/extension rates for three typical
pastes together with their strengths plotted as a
yield stress.

Another test described by Block was intended to
measure the 'hesion' of the pastes, although the extensional behaviour is probably being measured again.
Basically a disc is brought into complete contact with
the surface of a large volume of toothpaste. The disc
is then lifted away from the paste at different speeds
and the force produced is monitored. This produces a
similar flow field to the inverted syphon or "Fano"
flow experiment [116].

Initially there is a linear increase of force with
distance moved, but thereafter the forces maximise and
then decrease to a minimum, due to thinning of the

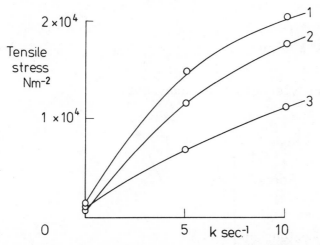

Fig. 23. Data from ref. 115 (table 1) replotted to give tensile
stress/average extension rate for curve:
1 : Dentifrice 2, 0.75% S C M C
2 : 70% sorbitol, 20% silica gel
3 : 70% sorbitol, 0.5% Carboxy vinyl polymer.

(draining) fluid column. The column of liquid that
can be supported until break depends on the 'stringiness' of the particular paste.

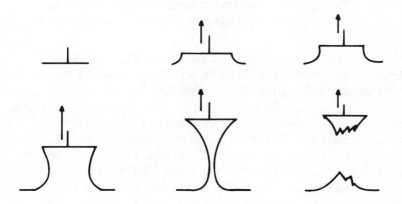

The results show that for a toothpaste 'base', the
"modulus" i.e. the force at fixed (small) distance is

a function of velocity (see Fig. 24) in such a way
that

$$\frac{Force}{Ext} = A + BV ,\qquad\qquad (17)$$

indicating viscoelastic behaviour.

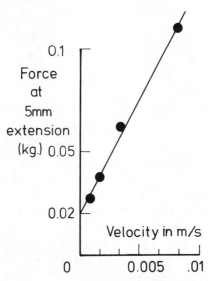

Fig. 24. Force is generated by lifting a cylinder of
toothpaste 5mm from a reservoir. Data replotted
from ref. 115 fig. 11. (Carragheenan A in sorbitol syrup).

This simple behaviour broke down at a critical
stress that varied slightly with increasing velocity
and was in the range 1000-1500 Nm^{-2}.

For fully formulated toothpastes, the behaviour is
not so simple. The force is not such a strong function of velocity; below 50 cm/min it is virtually
independent of velocity, so that the behaviour is
more solid-like. In this case the maximum force
decreases with increasing rate and is much higher.

Thus, whereas viscous forces are important in the
base, elastic forces dominate in the fully formulated
product. The critical extensions at which non-linear-
ity sets in are however similar for both materials.

3.7 Extremely viscous materials

3.7.1 Soap and non-soap detergents (NSD)

Soap, although normally considered a solid, can under extreme conditions (i.e. very long time or very high stress) behave as a liquid in that it is able to sustain steady flow. Soap has been known for many centuries as either hard (sodium stearate) or soft (potassium stearate), the former being the most convenient because it can be moulded or cut into bars. Non-Soap Detergents (NSD's) are simply the equivalent of compressed detergent powders, useful in primitive washing conditions (e.g. rivers) where they can be held and rubbed onto the soiled article, in much the same way as laundry soap was used in the U.K. early in the century. Their rheology is qualitatively similar to soap.

Linear testing

Linear testing of a range of commercially available soaps has been reported by Pacor et al. [117], using creep, constant shear rate and dynamic testing conditions. The range of experimental conditions were:

Creep test: Time $0-2 \times 10^4$ secs
     Shortest time reading $-$ 2 secs
     Typical stress range $- 0-4 \times 10^4 Nm^{-2}$
     Typical strain range $- 0-2 \times 10^{-3}$

Constant shear rate: Shear rate range
     $8 \times 10^{-6} - 1.6 \times 10^{-2}$ sec^{-1}
     Typical stress range $- 0-10^5$ Nm^{-2}

Dynamic testing: Frequency range:
     Free oscillation $- 0.8-40$ Hz
     Forced oscillation
     in bending mode
     [118] $- 200-300$ Hz

The data obtained gave information on the effect of water content, temperature, formulation and processing-induced orientation on the rheology.

Soap showed linearity below strains of approximately 10^{-3}. Typical values of the short time (5 sec) elastic modulus G' were found to be in the range $10^7 - 10^8$ Nm^{-2}, depending strongly on water content. The modulus doubled for about every 4% decrease in water content. The steady state viscosities varied over the range 5×10^{11} Nsm^{-2} to 100×10^{11} Nsm^{-2}, again doubling for every 4% water loss.

Dynamic measurements and inverted creep data correlated well to give values of G' in the frequency range 10^{-3} to 5×10^3 Hz, and values of $\tan \delta$ ($= G''/G'$) in the frequency range 2×10^{-1} to 10^3 Hz.

Typical data (replotted as log G' vs log freq) are shown in Fig. 25. The authors noted the general similarity between soap rheology and that of dispersions of tristearate crystals in paraffin oil. Similarly we would expect similar behaviour in greases (see Chapter 3), where soap crystallites in oil form an entangled network.

The microstructure responsible for the macroscopic rheology is probably a mass of intertwined soap crystals in the form of long ribbons, in a liquid-like continuum of electrolyte, water and lower molecular weight soluble soap fractions. Maclennan [119] carried out the early microscopic examination of soap using polarised light, and showed that molten soap solidifies on cooling to form what he called 'fibres'. The rate of cooling had a profound influence on the formation of these 'fibres'; rapid cooling produced irregular lumps of soap crystals together with very fine and very irregular fibres permeating the structure. Slow cooling on the other hand produced more fibres and more homogeneous crystal shapes.

In rheological terms no permanent cross links are present, so if stress is applied for long enough, steady flow eventually occurs ($\sim 10^6$ secs). Thus the value of $\tan \delta$ at very low frequencies will increase (see Fig. 26), showing that viscous behaviour dominates at long times (low frequencies).

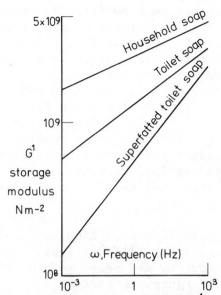

Fig.25. Data from ref.117 replotted as log G¹/log ω (curves represent solid lines from original G¹/log ω curve).

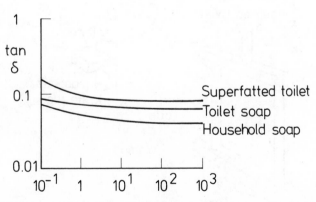

Fig.26. Values of tan δ vs. frequency for soaps illustrated in figure 25.

Data replotted from ref. 117.

Work at this author's laboratory has been carried out on the linear properties of soap in extension. The Deer Rheometer has been modified to measure the extensional properties by simple bending tests carried out at very small deformation. Fig. 27 shows the modified test geometry. The torque is applied to two vertically mounted rods of soap (obtained by extrusion from an orifice viscometer). The soap rods are clamped at the bottom and capped at the top with short soft iron sleeves. The torque is applied to the rods via the sleeves by magnetic knife edges mounted on radial arms. The knife edges ensure that no twisting of the rods occurs during bending, as would happen with normal clamping; a torsional correction would then be required.

The extensional compliance is then calculated from simple bending beam theory [120]. Typical angular displacement/time curves are shown in Fig. 28 for a number of applied torques. Departure from linearity is only noticed at the highest torque ($\sim$ 50 gm cm).

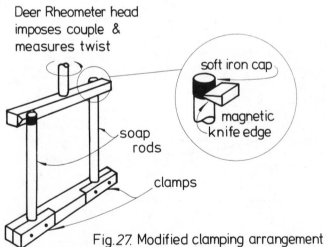

Deer Rheometer head imposes couple & measures twist

soft iron cap

magnetic knife edge

soap rods

clamps

Fig.27. Modified clamping arrangement to perform pure bending experiments on the Deer Rheometer (A similar arrangement can be used for the Weissenberg Rheogoniometer by driving the bottom clamps).

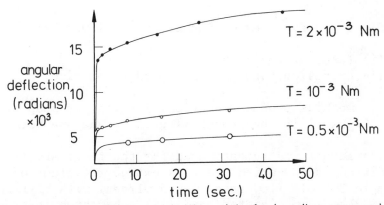

Fig.28. Angular displacement/time data for bending soap rods on the modified Deer Rheometer, with variable applied torque. (Note nonlinearity of highest torque data)

The same apparatus is used to measure the shear creep compliance by clamping both ends of a single soap rod in the middle of the apparatus.

Because the compliance only changes slowly with time, the data may be inverted to give E', G' vs ω curves, where E' is the extensional storage modulus.

For isotropic materials, E'/G' ≈ 3, but because soap has an alignable microstructure, orientation effects produce values of E'/G' > 6 for rods formed by extrusion through small (< 5 mm) orifices. This phenomena is well known in the polymer industry, and has been considered recently by van den Tempel in relation to general industrial processes [121].

Above deformations of 10^{-3}, soap shows non-linear behaviour. If simple pulling/compressing tests are carried out on an Instron type tester, the stress/ strain curve shows a continually decreasing slope with increasing deformation, until the stress is almost constant with increase in strain; the soap then fails.

Higher rates of extension can be achieved by:

(i) Penetration by sharp edged discs,
(ii) Extrusion through a sharp edged orifice.

These methods produce identical results for soap if the Cogswell [8] treatment is used to analyse both.

The actual instruments used for the tests are (i) a modified Maclow-Smith plastics extruder (volume of sample 300 ml, flow rates 0-6.5 ml/sec) with orifices from 2 to 7 mm diameter, and (ii) a benchtop Instron Tensile tester used in compression (via a 'cage' arrangement) with a range of sharp edged discs (d $\sim$ 3-10 mm) over a speed range 0.5 - 1000 mm/min.

Typical stress/extension rate curves measured by extrusion and disc penetrometry are shown in Fig. 29 together with the low k rod pulling/compressing data.

The behaviour in extension can be described for the steady state region by:

$$p_{11} = p_{11}^{(0)} + Kk^n ,$$ (18)

where $p_{11}^{(0)}$ is the yield stress.

$p_{11}^{(0)}$, K and n are very formulation and temperature dependent, but for fresh superfatted toilet soap at room temperature (water content $\sim$10%) values are approximately:

$$p_{11}^{(0)} \sim 5 \times 10 \ Nm^{-2} ,$$
$$K \sim 3 \times 10 \ Nm^{-2}s^n ,$$
$$n \sim 0.2 .$$

These variations in rheological properties caused by formulation variations have been noted as processing problems [122], so that soap which comes out of the extruder too soft (i.e. $p_{11}^{(0)}$ too low) is accounted for

by insufficient drying. This can cause the bars produced from the extrudate to be easily damaged in subsequent handling. Too little moisture in the soap also
causes problems, with slow and irregular extrusion.

A good description of the importance of soap rheology
is given in reference [123], under the title "Some
Toilet Soap Processing Faults", where soap rheology is
termed 'plasticity'.

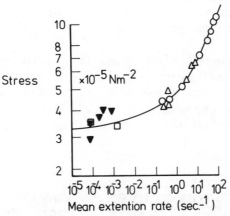

FIG.29. Steady state stress/mean extension rate
data on production toilet soap for the following
type of deformation:
- ▼ Steady pulling
- □ Steady compression
- △ Disc penetrometry (Sharp edged discs)
- o Extrusion through sharp edged orifices.

Other methods

Other (less quantitative) methods have been used to
measure soap rheology. The most widely used has been
sectilometry or cutting the soap with a wire. The
wire is either pulled through the soap and the force
produced is monitored, or otherwise the wire is

weighted and the velocity measured. The latter method
is usually employed to measure the yield stress by
measuring the minimum force needed to move the wire.

Bowen and Thomas [124] described a sectilometer to
measure what they termed the 'hardness' of soap. A
soap bar was driven onto a fixed wire at a constant
speed. The force on the wire was measured by the deflection of a strong spring. The speed of movement
was approximately 250 mm/min, but unfortunately the
wire dimensions were not given, but examination of the
published photograph of the apparatus shows the wire
diameter to be approximately 1 mm and the width of the
soap bar of the order of 50 mm. The force on the wire
was usually in the range 0-4 kg.

The effect of formulation (e.g. the ratio of palm
oil to tallow), cooling mode and electrolyte level
(for a variety of salts) was measured. Typical
'hardness' values are shown in Fig. 30 as a function
of stated formulation and process variables.

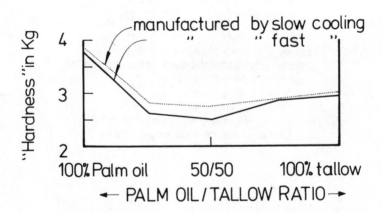

FIG.30 Wire cutting (sectilometry) data on soap
produced by slow or rapid cooling of a
range of formulations.

Data taken from ref. 124.

We have tried in the author's laboratory to make
this test more quantitative and have defined average
stresses and (by analogy with the disc penetrometer)
strain rates:

If we define $\bar{p}_{11} = K_1 \dfrac{F}{2\ell R}$ Nm^{-2} (19)

and $\qquad\qquad \bar{k} = K_2 \dfrac{2V}{R}$,   (20)

where ℓ is the length of wire cutting the soap,
$\qquad$ R is the radius of wire cutting the soap,
$\qquad$ V is the velocity of wire cutting the soap,
$\qquad$ F is the force exerted on the wire,

then measurements on previously characterised soap
bars show that $K_1 \approx K_2 \approx 1$. Fig. 31 shows the values
of $\bar{p}_{11}$ and $\bar{k}$ found by this method compared with those
from extrusion and disc penetrometry. The sectilo-
meter was a frame mounted on the Instron Tensile
Tester.

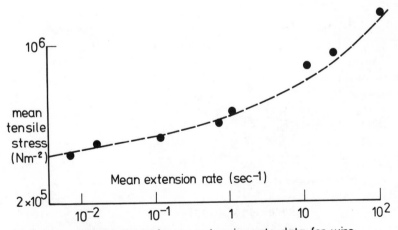

Fig. 31. Mean tensile stress/mean extension rate data for wire
$\qquad$ experiments (sectilometry) on a typical toilet soap.

$\bullet$ data
-----mean curve of equivalent data from fig. 29.

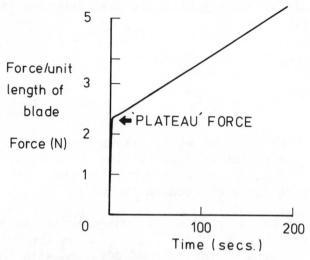

Fig. 32. Typical `knife´ cutting curve:

Cutting force vs. time. Knife width 0.0625 mm, blade length 28mm cutting speed 8mm/min.

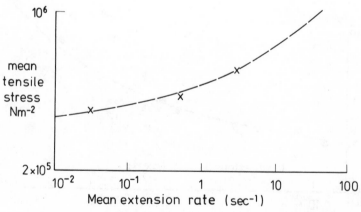

Fig. 33. Mean tensile stress/mean extension rate data for a typical toilet soap calculated from initial plateau values of knife cutting experiment.

   x   Data

--- Mean curve of equivalent data from figs. 29 & 31.

Similar experiments were performed using steel feeler gauges for cutting the soap. Unlike the wire experiment, where the force was constant with penetration, the force measured on the feeler gauges reached a plateau quite quickly, but then slowly increased with depth due to the increasing friction on the blade from the soap, see Fig. 32. If the initial plateau value is used, the same expressions may be used for stress/strain rate as for the wire (with the thickness of the gauge = 2R). Again good agreement is found with other experiments (see Fig. 33).

The measurement of 'yield values' of commercial soap using the minimum force necessary to move the wire was reported by Vold and Lyon [125], using a cutting wire radius of 0.02 cms. The criteria of "just beginning to move" meant for them a velocity not greater than 1 mm/5 minutes (i.e. using the above equation k < 0.03 sec^{-1}). From our experience (see Fig. 31) this would give a stress slightly higher than a true yield stress and in some situations would lead to a significant contribution from Kk^n.

Typical curves obtained for "yield stress" as a function of temperature and water content are shown in Fig. 34. Their findings relating yield stress to processing history were that "the slowest cooled soap was the softest and the fastest cooled (quenched) soap was the hardest". This agrees with the findings of Bowen and Thomas [124]. They found, however, little correlation between the yield value and the supposed phases present in the soap.

Penetrometry is also used to assess the rheological properties of soap. A weighted truncated pin is brought into contact with the surface of the soap bar, and is then allowed to penetrate the soap. The penetration is monitored as a function of time. Eventually enough of the pin enters the soap for the stress to equal the yield stress, and the pin stops (see Fig. 35). Again an estimate of the stress/strain rates can be made but this time the frictional component cannot be eliminated. Thus the data is best left as measured. The 'plasticity' of soap is such that standard methods used for penetrometry of petroleum waxes and bitumen are most suitable (i.e. ASTM D1321-76, BS 4691 : 1971).

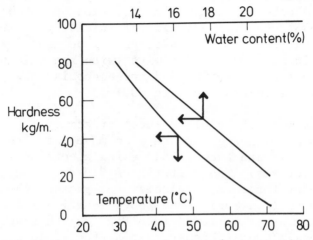

Fig.34. Typical dependence of hardness' measured by sectilometry on temperature and water content.
Data taken from ref.125;

On-line Measurements

On-line measurement of viscosity has a useful part to play in the actual manufacture of soap*. For many years soapmaking was carried out as a batch operation, but continuous processing has recently become available. One problem encountered was in finding some way of monitoring the degree of saponification. Alfa-Laval introduced a method of automatic control which used the (known) variation of soap viscosity with caustic soda concentration (see Fig. 36). This method involved the measurement of the differential pressure across a circulation pump in the saponification circuit. The viscosity curve shows a well-defined maximum at about 0.2% excess NaOH, falling off rapidly if

---

* For a general description of toilet soap manufacture see reference [122].

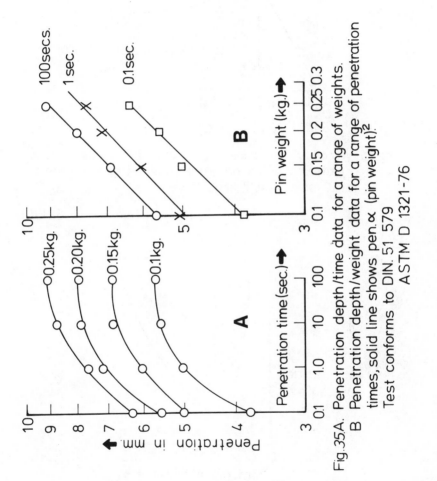

Fig.35A. Penetration depth/time data for a range of weights.
B Penetration depth/weight data for a range of penetration
times, solid line shows pen.∝ (pin weight)^2
Test conforms to DIN. 51 579
ASTM D 1321-76

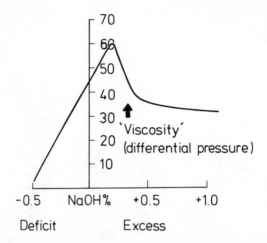

Fig.36 'Viscosity' (measured as differential pressure across a metering pump: arbitrary units) vs. caustic soda level for a typical soap.

(Taken from ref. 126)

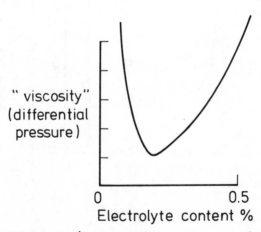

Fig.37. Viscosity (measured as pressure drop across a metering pump: arbitrary units) vs. electrolyte content for a typical soap.

(taken from ref. 126)

there is a deficit of electrolyte and almost as rapidly in the case of an excess. Since this rheological behaviour is well defined, it is used to control the rate of the caustic proportioning pump automatically [126,127]. It is claimed that this form of control is "highly accurate, robust and requires virtually no maintenance".

Soap "fitting" (i.e. obtaining the correct soap phase via electrolyte control) is also achieved by in-line viscosity monitoring and control. Here the viscosity curve shows a minimum with electrolyte concentration [53], see Fig. 37. Control is restricted to the right hand branch of the U-curve, where the viscosity (again monitored via the pressure drop across a pump) is used to control the electrolyte proportioning pump. Since the viscosity is very temperature sensitive, this method of control can only be used where the saponification and fitting operations are carried out at constant temperature.

4. RHEOMETRY AND PATENTING

In a survey of 4000 recent patents filed in the detergents and shampoo area, 185 made mention of rheological features; of these only 27 quoted any measured values; of these only two quote the actual conditions of measurement of viscosity at stated shear rates [125]. Two patents quote yield stress, but these were probably extrapolated from shear rates available on normal commercially available viscometers [129].

Typical examples of undefined measurements are:

(i) "Floor washing detergent - its viscosity is low - 30 poise at 25°C" [130].
(ii) A thinner for sodium alkyl sulfate (a detergent raw material); 2% polyethylene glycol (MW = 1500) reduces the detergent viscosity from 90,000 cp to 619 cp at 27°C, compared with 5,800 cp using $MeC_6H_4SO_3Na$ [131].
(iii) "A liquid washing composition with viscosity 100 - 2,000 cp" [132].

(iv) "A liquid scourer, 7,000-30,000 cp, preferably
7,000-25,000 cp" [133].

All these examples contain information about liquids
that are non-Newtonian. The viscosities quoted are
without stated shear rates, are thus ambiguous, and
can form no real legal definition of the liquid!

Usually, rheological properties are only given verbal descriptions, with no viscosity figures at all:

"pour point, non-gelling, fluidity, pourable, jelly-
like, thick, pumpable, plastic consistency".

Sometimes a verbal description is supplemented by a
viscosity figure (usually undefined) such as:

"viscosity < 100 cp, thus easy to transport, handle
and use [134]; previously such high concentrations
were of a pasty consistency".

"controlled viscosity liquid detergent 40-120 cp at
24°C and fluid to 70°C" [135].

"good viscosity properties - 75-200 cp at room
temperature" [136].

"Hard Surface fluid detergent composition, yield
value 5-600 and apparent viscosity < 12,000 cp"
[137].

"Oven cleaner, viscous liquid - 800-10,000 (no units
mentioned)".

"Shampoo - high viscosity, > 4,000 cp at 23-250°C"
[138].

Care is not often taken to define the rheometer used,
or the operating conditions.

Proper rheometry would remove the subjective element
in rheological description. For instance, when one
speaks of a "liquid" sufficiently fluid to pour readily [139] this can be subject to a considerable breadth

of interpretation, whereas a proper viscosity measurement would fix the range completely. One would expect a carefully worded legal document to contain equally carefully defined rheology, at least as carefully defined as the chemistry quoted in patents.

As a company's strength is often measured by its patents, detergent companies in general seem very weak in rheological characterisation.

5. CONCLUSION

While there have been reviews of other industries' use of rheology and rheometry [140-144], little seems to have been published to correlate the data available from the detergents industry. However, from what we have considered it is clear that rheometry has much to offer if carried out correctly and also if done in conjunction with other physical chemistry techniques (e.g. NMR) in basic studies.

As we have seen, the basic requirements of rheometry are:

(i) Correct simulation of flow deformation fields, e.g. extensional vs shear [121].
(ii) Correct data processing to convert force/ deformation data into material parameters via geometrical factors.

These are difficult requirements to meet in simple flow systems — thus we have the paradox that the simplest experiments to perform (e.g. the Ford cup) are the most difficult to analyse. However, progress is being made and in future the industry will be able to gain from the correct application of appropriate rheometry.

110

REFERENCES : CHAPTER 2

[1] A.K. Reng and W. Skrypzak, Cosmetics & Toilet-
     ries 94 (1979) 25

[2] J. Sestak and F. Ambros, Rheol. Acta 12 (1973)
     70

[3] O. Wein, Rheol. Acta 16 (1977) 248 and J. Non-
     Newtonian Fluid Mechanics 1 (1976) 357

[4] K. Wichterle et al., Chem. Prom. 27(9) (1977)
     462

[5] H. Hoffman, Tenside Deterg. 14(3) (1977) 134

[6] J.R. Van Wazer et al. "Viscosity and Flow
     Measurement", Wiley Interscience, New York,
     1963

[7] E.C. Bingham, "Fluidity and Plasticity", McGraw
     Hill, New York, 1st Edition 1922, p.328

[8] F.N. Cogswell, J. Non-Newtonian Fluid Mechanics,
     4 (1978) 23

[9] "Cationic Surfactants", ed. E. Jungermann,
     Marcel Dekker, New York, 1970, p.253

[10] J.G. Kirkwood and P.L. Auer, J. Chem. Phys. 19
     (1951) 281

[11] S. Gravsholt and H. Bjoernoel. Proc. Int.
     Congr. Rheol., (7th) 1976, p. 300

[12] N. Pilpel, Trans. Farad. Soc. 62 (1966) 2941

[13] F.E. Bailey and J.V. Koleske, "Nonionic
     Surfactants", p.801

[14] Anon. S.P.C. p.478, June 1966

[15] G.D. Smith, JAOCS 56(2) (1979) 87

[16] G. Barker et al., Soap, Cosmetics, Chemical Specialities, March 1978, p.38

[17] Anon. Manuf. Chemist and Aerosol News, p.29, March 1979

[18] Anon. "Which?", p.100, April 1973

[19] P. Sherman, SPC, p.54, Jan. 1972

[20] J.W. Hoyt, Trans. AIME (J. Basic Engineering) 94D (1972) 258

[21] J.L. Zakin and J.L. Chang. Int. Conf. on Drag Reduction, Cambridge, 4-6 September 1974, Organised by BHRA.

[22] N.S. Berman, Annual Rev. Fluid Mech., Vol.10, 1978

[23] J.G. Savins, U.S. Patent No. 3,361,213 (1968)

[24] A. White, Nature (Phys. Sci.) 235 (1972) 154

[25] W.K. Lee et al., A.I.Ch.E.J. 20 (1974) 128

[26] A.B. Metzner, Phys. Fluids, 20, 145, 1977

[27] C. Barker et al., J.C.S. Faraday I, 70 (1974) 154

[28] A.V. Shenoy, Rheol. Acta 15 (1976) 658

[29] P. Sherman, "Emulsion Science", Academic Press, New York, 1968

[30] P. Sherman, "Industrial Rheology", Academic Press, London, 1970

[31] Anon. Cosmetics & Toiletries, 93, July 1978

[32] N. Ohba, Bull. Chem. Soc. Japan, 35 (1962) 1175

[33] H. Komatsu and M. Takahashi, Trans. Soc. Rheol. 21 (1977) 219

112

[34] B.H. Zimm and D.M. Crothers, Proc. Nat. Acad. Sci., U.S.A., 48 (1962) 405

[35] D.W. de Bruijne, Rheol. Acta 13(3) (1974) 418

[36] M. van den Tempel, J. Coll. Sci. 16 (1961) 284

[37] J.M.P. Papenhuijzen, Rheol. Acta 11 (1972) 73

[38] J. Lyklema et al., Rheol. Acta 17 (1978) 525

[39] H. Beltman, Commun. Agric. Univ. Wageningen, Neth. 74 (1974) 5

[40] H.R. Moskowitz and D. Fishken, Cosmetics & Toiletries 93 (1978) 31

[41] E.L. Roehl, S.P.C. p.343, 1972

[42] B.W. Barry and G.M. Eccleston, J. Text. Studies 4 (1973) 53

[43] G.M. Eccleston, Cosmetics & Toiletries 92 (1977) 21

[44] P. Jansson and S. Friberg, Mol. Cryst. Liq. Cryst. 34 (letter) (1976) 75

[45] Anon. S.P.C. p.50, February 1976

[46] H.W. Bücking and A. May. Ciba-Geigy Monograph "Household Detergents", February 1976, p.34

[47] Uiber Kationische Wäsheveredler, Sch. Witzer, M.K. London F.S.A. 70 (1968) 349

[48] A.D.C. James and P.H. Ogden, JAOCS 56 (1979)

[49] G. Scott-Blair, personal communication.

[50] Anon. S.P.C. p.50, February 1976

[51] G. Noirel and J. Renaud, Ciba-Geigy Monograph "Household Detergents", February 1976, p.28

[52] A.G. Dodson, Ph.D. thesis, University of Wales, 1973

[53] D.F. Farrow, J. Chem. Soc. 101 (1912) 347

[54] Bundenberg de Jong et al., Proc. Koninkl. Nederland Akad. Wetensh 51 (1948) 1179, 52 (1949) 3

[55] J. Winsor, J. Coll. Sci. 10 (1955) 88

[56] N.J. Pilpel, Phys. Chem. 60 (1956) 779

[57] N.J. Pilpel, Trans. Farad. Soc. 62 (1966) 1015

[58] J.G. Oldroyd, Proc. Roy. Soc. A218 (1953) 122

[59] A.S. Lodge, "Elastic Liquids", Academic Press, 1964, Chapters 4 and 6.

[60] P. Debye and E.W. Anacker, J. Phys. Chem. 55 (1951) 644

[61] J.L. den Otter and J.M.P. Papenhuijzen, Rheol. Acta 10(1) (1971) 457

[62] H. Janeschitz-Kriegl and J.M.P. Papenhuijzen, Rheol. Acta 10(4) (1971) 461

[63] G.J.T. Tiddy and P.A. Wheeler, Jour. de Phys. Colloque. CI, Supple. 3, 36, March 1975, C1-167

[64] D. Saul et al., J. Chem. Soc. Farad. Trans. I, 70 (1974) 163

[65] D.C. Jones, J. Chem. Soc. (1951) 126

[66] A.C. Zettlemayer, J. Coll. Int. Sci. 28 (1968) 343

[67] A.J. Hyde and D.W.M. Johnstone, Proc. Int. Congr. Rheol. (7th) (1976) p.564

[68] H.A. Barnes et al., Rheol. Acta 14 (1975) 53

[69] H.A. Barnes et al., Rheol. Acta 14 (1975) 61

[70] P.E. Rouse, J. Chem. Phys. 21 (1953) 1272

[71] W.W. Graessley, J. Chem. Phys. 54(12) (1971) 5143

[72] B.D. Marsh and J.R.A. Pearson, Rheol. Acta 7 (1968) 326

[73] M. Doi and S.F. Edwards, J.C.S. Faraday II, 74 (1978) 560

[74] M. Doi and S.F. Edwards, J.C.S. Faraday II, 74 (1978) 918

[75] H.C. Boiij, Ph.D. Thesis, Leiden, 1970.

[76] F.E. Bailey et al., Ind. Eng. Chem. 58 (1958) 8

[77] W.K. Asbeck, J. Soc. Cosmetic Chem. 8 (1957) 282

[78] J.T. Bergen and W. Patterson, J. Appl. Phys. 24 (1953) 712

[79] F. Rodriguez and L.A. Goettler, Trans. Soc. Rheol. 8 (1964) 3

[80] W.K. Asbeck and E. Baxter, American Chem. Soc. Meeting, Polymer Div. Chicago, September 1958

[81] F.E. Bailey et al., "Nonionic Surfactants", ed. M. Schick, Arnold, London, 1967, Chapter 23.

[82] G. Boardman and R.L. Whitmore, Nature, 187 (1960) 50

[83] G.J.T. Tiddy, Personal communication.

[84] M. Kleman et al., Adv. Chem. Ser. 57 (1976) 362

[85] P. Solyom and P. Ekwall, Rheol. Acta, 8 (1969) 316

[86] B. Hallstrom and S. Friberg, Acta Pharm. Suec,
      7 (1970) 691

[87] B. Tamamushi and M. Matsumoto, "Liquid Crystals
      and Ordered Fluids" II, Plenum Press, 1974,
      p.711

[88] B. Tamamushi et al., Colloid Polymer Sci. 254
      (1976) 571

[89] M.J. Groves and A.B. Ahmed, Rheol. Acta, 15
      (1976) 501

[90] R.W. Duke and L.L. Chapoy, Rheol. Acta 15 (1976)
      548

[91] R. Horn and M. Kleman, Journal de Phys., Special
      Issue, December 1978, p.250

[92] R. Horn and M. Kleman, Annal. Phys. 2-3-4

[93] F.J. Kahn, Appl. Phys. Letters, 22 (1973) 386

[94] J.R. Fisher and A.G. Frederickson, Mol. Cryst.
      Liq. Cryst. 8 (1965) 267

[95] A.R. White et al., Mol. Cryst. Liq. Cryst. 43
      (1977) 13

[96] L. Bohlin and K. Fontell, J. Colloid and Inter-
      face Sci., 67 (1978) 272

[97] S. Gravsholt, Polymer Colloids II, ed. R. Fitch,
      Plenum Press, New York, 1979

[98] A.J. Hyde et al., Proc. VIth Inter. Congr.
      Surface Active Subs., Zurich 1972, Vol. II(2),
      813-823

[99] E. Bose, Physik. -Z. 8 (1907) 347

[100] S.H. Bastow and F.P. Bowden, Proc. Roy. Soc. 151
      (1935) 220

[101] W. Ostwald and H. Malss, Kolloid Z. <u>63</u> (1933) 192

[102] S. Glasstone, "Textbook of Physical Chemistry", 2nd ed. p.515, Van Nostrand, Princetown, N.J., 1946

[103] F.W. Southam and I.E. Puddington, Can. J. Res. <u>B-25</u> (1947) 125

[104] B.D. Powell and I.E. Puddington, Can. J. Chem. <u>31</u> (1953) 828

[105] R.S. Porter and J.F. Johnson, "Rheology", ed. F.R. Eirich, Academic Press, New York and London, 1967, Vol. 4, Chapter 5.

[106] H. Stache, 'Seifen Industrie Kalendar', 1969, p.87, Klasing, Berlin. See also Anon. S.P.C. November 1970, p.708

[107] Anon. S.P.C. <u>43</u> (1970) 708

[108] A.B. Metzner, "Advances in Chem. Eng.", Vol. I, Academic Press, New York, 1956, p.79

[109] F.A. Holland and F.S. Chapman, "Liquid Mixing and Processing", Reinhold, New York, 1966

[110] Y.S. Su and F.A. Holland, Chem. and Proc. Eng., September 1968

[111] P.H. Calderbank and M.B. Moo-Young, Trans. Inst. Chem. Eng. <u>39</u> (1961) 337

[112] M.F. Edwards and W.L. Wilkinson, The Chem. Engnr., September 1972, p.328

[113] G. Astarita and R.A. Mashelkar, The Chem. Eng., Feb. 1977, p.100

[114] A.H.P. Skelland, "Non-Newtonian Flow and Heat Transfer", John Wiley, New York, 1967

[115] M. Block, J. Soc. Cosmetic Chem. 26 (1975) 189

[116] G. Astarita and L. Nicodemo, Chem. Eng. J. 1 (1970) 455

[117] P. Pacor et al., Rheol. Acta 9(3) (1970) 455

[118] C.J. Nederveen and C.W. van der Wal, Rheol. Acta 6 (1967) 316

[119] J. Maclennan, JSCI, 5th October 1923

[120] A. Nadai, "Theory of Flow and Fracture of Solids, McGraw Hill, New York, 1950

[121] M. van den Tempel, The Chem. Engnr., February 1977, p.95

[122] Anon. S.P.C. December 1970, p.790

[123] Anon. S.P.C. November 1972, p.683

[124] J.L. Bowen and R. Thomas, Trans. Farad. Soc. 31 (1935) 164

[125] R.D. Vold and L.L. Lyon, Ind. Eng. Chem. 17 (1945) 585

[126] G.R. Platt, S.P.C. April 1970, p.233

[127] Anon. S.P.C. December 1970, p.787. Reprint of Technical Bulletin, Haarmann & Reimer GmbH, Holzminder, West Germany.

[128] Unilever Patents, BE 809,955 and GB 31,090/70

[129] Patent Number GB 1,167,597 and GB 1,181,607

[130] Patent Number SU 462,861

[131] Patent Number JA 50/116,383

[132] Patent Number NL 7,413,265

118

[133] Patent Number BL 605,230

[134] Patent Number DT Appl. No. 2,501,982

[135] Patent Number US 536,655

[136] Patent Number US 3,634,268

[137] Patent Number SW 11,916/69

[138] Patent Number NL 7,612,151

[139] Patent Number UK 1488 352

[140] Paint Industry: S. Bryan, Pigment and Resin
       Tech., August 1977, p.7

[141] Paint Industry: S. Le Sota, Paint and Varnish
       Prod., 47 (1957) 60

[142] Oil Industry: F.J. Villforth et al.,
       Lubrication, 56(2) (1970) 25

[143] Cosmetics Industry: A.L. Scarbrough, J. Soc.
       Cosmetic Chem. 8(5) (1957) 306

[144] K. Walters, Rheometry, Chapman & Hall, 1975.

CHAPTER 3

LUBRICANTS

By

J.F. Hutton

CONTENTS

1. INTRODUCTION

It is conventional in lubrication technology to divide the operating regimes into the three classes: hydrodynamic, boundary and dry. In hydrodynamic lubrication the bearing surfaces are held apart by pressures generated in a fluid film and wear is nil. As the effectiveness of hydrodynamics diminishes, and the surfaces come closer together, high-spots on one of the surfaces hit high-spots on the other and wear begins to increase. This is the boundary lubrication regime in which success or failure depends a great deal on the formation and properties of surface films, which are usually laid down by adsorption or chemical reaction of additives. In very severe conditions, by accident or design, the bearing contact may be dry in the sense that no liquid lubricant (and in outer space, no fluid lubricant of any sort) is present. In dry lubrication everything depends on the properties of the bearing materials and surface films, whether these are formed *in situ* or applied.

Hydrodynamic lubrication can be divided into two important sub-classes: classical hydrodynamic lubrication and elastohydrodynamic lubrication. In classical hydrodynamic theory the bearing surfaces are considered to be rigid and undeformable, whereas in elastohydrodynamics they are assumed to deform elastically under the influence of hydrodynamic and applied

forces. As a rule classical hydrodynamic lubrication
occurs between pairs of conforming surfaces, that is,
when the radius vectors at the contact have the same
sign. Hence, for example, journal bearings operate
in this regime. For counterformal surfaces, that is,
when the radius vectors at the contact have opposite
signs, the contact pressures are almost unavoidably
so high that elastic deformations occur and the state
of lubrication is elastohydrodynamic. Ball- and
roller-bearings and gears operate in this regime.

Another classifying term, which is not in widespread
use but is relevant in the present context, is rheodynamic lubrication. It has no formal definition, but
it covers the whole of the thick oil-film range with
explicit recognition of the fact that the lubricant
fluid under consideration is displaying non-Newtonian
behaviour.

Fluid rheology and rheometry are of prime importance
in the hydrodynamic and much of the boundary regimes,
but irrelevant in dry lubrication. Therefore, in this
chapter we shall be concerned with the rheometry
relevant to continuous fluid layers. It also simpli-
fies matters if in this chapter we consider lubricants
to be either mineral oils or greases. In practice, a
variety of fluids are used as lubricants, including
gases. Process liquids of all sorts are often
required to lubricate the pumps and metering devices
through which they flow. For example, liquid fuels
ranging from low-viscosity aircraft fuels to very-
viscous industrial fuels must have a lubricating
function. Similarly with hydraulic fluids: these are
liquids that include water-in-oil emulsions and oil-in-
water emulsions as well as mineral oils and aqueous
solutions. It would occupy too much space and would
involve a degree of repetition if all such liquids
were included. We rely on the reasonable assumption
that the hydrodynamical principles are the same whatever the fluid and on the fact that the rheometers,
where they are used, are those used for mineral oils
and greases.

2. RHEOMETRY OF LUBRICATING OILS

2.1 Nature of lubricating oils

The lubricating oils under consideration originate from the distillation of crude oil. The primary distillation is carried out at atmospheric pressure to remove the lighter gasoline, kerosine and gas oil fractions, and the lubricating oils are present in the residue that boils above about 370°C. This residue is further distilled under high-vacuum conditions to give a number of lubricating-oil fractions and an asphaltic bitumen residue. This second stage residue from crudes low in asphalt is also a source of lubricating oils, and these are known as residual oils to distinguish them from the so-called distillate oils that boil off in the high-vacuum still. Again, depending on the source of the crude, the lubricating-oil fractions constitute up to 5% of the total crude.

Following the distillation processes, the lubricating-oil streams are further refined by treatment with acid or solvents to remove unwanted components. By treatment with hydrogen at moderate temperatures and pressures, unsaturated hydrocarbons can be saturated and sulphur compounds can be removed. Paraffinic compounds which separate as waxy solids at low temperatures are removed, as required, by chilling and filtering or by a solvent treatment. A final treatment with clay removes polar compounds and dark colouring matter from the oil.

The refinery streams flowing from the above-mentioned processes are known as base oils. Base oils are characterised by their crude origin, refining process, viscosity, Viscosity Index* and, if required, by other more specific properties. Finished lubricants are blends of base oils with or without the addition of additives that confer special properties. Additives

---

* Viscosity Index (VI) is a measure of the variation of viscosity with temperature, the higher the VI the smaller the variation: ISO Method 2909: IP Method 226: ASTM Method 2270.

are usually supplied as concentrates in mineral oil
and they are readily added during the blending process.

Mineral oils are complicated mixtures of thousands
of types of organic molecules. Detailed analysis is
impossible, but certain chemical classes can be identified, e.g. paraffins and aromatics. This leads to
another way of classifying oils: by chemical class.
Additives make oils even more complicated chemically.
They are classified according to their function, and
include boundary-lubrication additives, extreme-
pressure lubrication additives, oxidation inhibitors,
additives to neutralise acids, and surface-active
materials to disperse, and maintain in suspension,
solid contaminant particles. Polymers are added to
oils to modify rheological properties. One type of
polymer, a VI improver, reduces the variation of viscosity with temperature. Another type, a pour-point
depressant, inhibits the formation of a wax-crystal
network at low temperatures and this extends the
liquid range of the lubricant.

We should mention in passing synthetic oils. These
are not mineral oils, by definition, but they are
widely used, although expensive. Chemically simpler
than mineral oils the main types are silicone fluids,
organic esters, polyalkylene glycol derivatives,
phosphate esters and polyphenyl ethers. A major
characteristic, as with mineral oils, is the wide
temperature range over which these materials remain
liquid.

2.2 Why rheometry?

2.2.1 Background to standardised methods

In classical hydrodynamic lubrication most problems
can be solved with the aid of the Reynolds equations,
which are approximations to the Navier-Stokes equations for incompressible, viscous, inelastic fluids.
Therefore, from a rheological point of view, a knowledge of the viscosity is sufficient to determine the
oil film thickness and the friction. However, since

oils are used over wide temperature ranges, and since
viscous heating may occur and thus affect the validity
of the isothermal solution of the Reynolds equation,
the viscosity-temperature function is important. It
is for these reasons that the viscosity and the variation of viscosity with temperature (by Viscosity
Index) have become the subjects of standardised
methods*. Glass capillary viscometers are used for
the measurements. For ease of comparison amongst and
between groups of workers, the temperatures at which
the low-shear viscosity is measured have been standardised at 40°C and 100°C. In a recently introduced
viscosity-grading system** the basis is the kinematic
viscosity at 40°C. In this system, for example, a
10-grade oil has a viscosity of between 9.0 and 11.0
cSt at 40°C. There are six grades for each decade in
viscosity.

An illustration of the amount by which the kinematic
viscosity of oils varies with temperature is given in
Figure 1 [1]. Results are given for two mineral base
oils and one synthetic fluid plotted in such a way
that the graphs are linearised for most mineral oils.
In the industry a VI of 101 is considered to be high
whilst a value of -19 is very low.

Some 30 years ago it was found that adding a small
amount of polymer to an oil reduced the rate of decrease in viscosity with increase in temperature, with
the result that many modern oils contain polymeric
additives. The rheometrical consequence of this is
that the viscosity-rate of shear function is now required to be known or, at least, characterised. The
types of polymer used include polyisobutene, polyalkyl-
methacrylates and the copolymers styrene-butadiene and
styrene-isoprene. Molecular weights from about 50 000
up to some hundreds of thousands are used. It is
difficult to talk about typical rates of shear, or

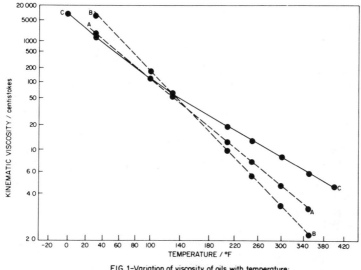

FIG. 1-Variation of viscosity of oils with temperature:
A. 101 VI mineral oil
B. -19 VI mineral oil
C. 186 VI polypropylene glycol

(Reproduced from C.M. Murphy, J.B. Romans and W.A. Zisman. Trans. ASME(1949)71 561)

shear stresses, but motor-oil specialists consider
that at the cranking speeds required to start a
gasoline engine the effective rate of shear in the
engine is about 10^4 s^{-1}. On the other hand, in normal
engine operation, rates of shear of 10^6 s^{-1}, at
temperatures between 100°C and 150°C, are experienced
in journal bearings and between a piston ring and the
cylinder wall. But much lower rates of shear are also
involved, particularly in starting an engine at low
temperatures, as will be discussed later. The industry
has taken notice of the importance of rate of shear
and has introduced a low-temperature, high-shear
standard test method, the Cold-Cranking Simulator*.
The industry is also actively considering the introduction of a low-temperature, low-shear test and a
high-temperature, high-shear test.

Figure 2 [2] shows the variation of viscosity with
rate of shear for an experimental 5W/20 multigrade oil
thickened with a methacrylate polymer. Measurements
were made with a capillary viscometer at 100°F, and

* The Cold-Cranking Simulator, ASTM Method D2602.

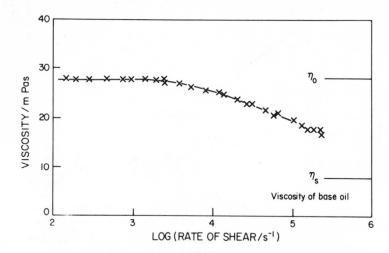

FIG. 2 – Capillary viscometer results, after correction, for a
multigrade oil containing methacrylate polymer

(Reproduced from W. Philippoff. ASLE Trans.(1958) 1 82)

the results were corrected for kinetic energy losses
and for the effect of pressure on viscosity.

The introduction of new test methods is a slow process. In the first place it may take 5 or 10 years to
establish the need for a new method. Then an official
body, such as the Institute of Petroleum (UK) or the
American Society for Testing and Materials (US), will
develop the methods, and this process may take a
similar period of time to complete. The choice of
test conditions, for example, temperature and rate of
shear, is generally made on the basis of known correlation between performance and laboratory measurement.
In making this choice, the relative merits of a number
of such correlations have to be judged. The choice of
instrument and operating procedure has to be made
bearing in mind the capital and operating costs involved, and balancing the lowest cost against the
highest accuracy and precision. It will often be
necessary for suppliers and users of lubricants, both
large and small companies, to make the measurements on
a routine basis, so the equipment must be within the
means of smaller firms and it must be capable of being
used by relatively unskilled staff.

However, the process does not stop with the introduction of a new method; the lubricant properties are
invariably incorporated in a specification. In the
case of low temperature rheological measurements of
motor oils the influential Society of Automotive
Engineers (US) instituted the correlation work, and
invited ASTM to develop laboratory methods, with the
object of modifying the SAE classification of engine
oil viscosity. The 1979 version of the classification
is given in Table 1. It is based on a low-shear
viscosity measurement at 100°C and a high-shear viscosity measurement at -18°C. As an example, a 20W/50
multigrade oil has a viscosity between 21.9 and 16.3
cSt at 100°C and between 2500 and 10 000 cP at -18°C.
The classification is used universally, hence changes
cannot be made without full consideration of the
consequences to both supplier and user.

Table 1

SAE viscosity grades for engine oils

SAE viscosity grade	Viscosity range		
	Centipoises (cP) at -18°C (ASTM D 2602)	Centistokes (cSt) at 100°C (ASTM D 445)	
	Max.	Min.	Max.
5W	1250	3.8	-
10W	2500	4.1	-
20W[a]	10 000	5.6	-
20	-	5.6	less than 9.3
30	-	9.3	less than 12.5
40	-	12.5	less than 16.3
50	-	16.3	less than 21.9

Note: 1 cP = 1 mPas; 1 cSt = 1 mm^2/s

[a] SAE 15W may be used to identify SAE 20W oils which have a maximum
viscosity at -18°C of 5000 cP.

Reproduced from "SAE Handbook"

Society of Automotive Engineers, Warrendale, Pa. 15096 (1979).

SAE Recommended Practice J300d

The use of the Cold-Cranking Simulator for the measurement at $-18^{\circ}C$ (originally the test temperature was $0^{\circ}F$) was introduced in 1967. It replaced what was in effect a low shear viscosity, which was obtained by extrapolating to a low temperature measurements made in glass capillary viscometers at higher temperatures. At the present time SAE are considering further changes; those at low temperature having reached a more advanced stage than those at high temperature. It is proposed that the low-temperature W-grades be specified in terms of the *temperature* at which certain values of rheological parameters are reached, and this has been agreed by many of the parties concerned. Two parameters are involved, namely, the high-shear Cold-Cranking Simulator (CCS) viscosity and a yield stress. The CCS viscosity value is likely to be 3.5 Pas and the yield stress 105 Pa. As regards the viscosity at high temperatures a new measurement at a high rate of shear is under consideration. Thus, the new proposals represent considerable changes with very good technical justification as will be shown later. Great care is being exercised to ensure that the introduction of the new concept is beneficial to all concerned and does not give rise to unnecessary complication and expense.

Following the above brief digression into the activities of standardising and specification bodies we return to the topic of liquid lubricants and the reasons why rheometry is needed.

Table 2 shows some representative values for conditions in US engines derived from data quoted by Fischl, Horowitz and Tutwiler [3]. Recent experiments have shown that a motor oil in European engines can experience still higher temperatures than those quoted. Thus, Bell and Voisey [4] showed that for engines run in severe conditions, in field tests as well as bench tests, there was a better correlation between journal bearing wear and viscosity measured at $150^{\circ}C$ and 10^{6} s^{-1} than viscosity measured at $100^{\circ}C$ and 10^{6} s^{-1}. This result and other observations indicate that $150^{\circ}C$ is relevant to bearing operation. On the other hand, Bell and Voisey found that piston-ring wear, under slightly less severe operating conditions, correlated

Table 2

Relative engine shear rates and stresses

Engine operation	Oil temperature, ^{o}C	Speed, rev/min	Shearing rate,[a] s^{-1}	Oil viscosity,[b] Pas	Shearing stress,[c] Pa
Running	100	3000	100×10^4	0.01	1×10^4
Cold starting	-18	30	1×10^4	2.5	2.5×10^4

[a] Average effective shear rate generally estimated for critical engine parts (bearings, cylinder walls, and so forth).

[b] Typical of 10W-30 oils.

[c] Equals product of shearing rate and viscosity.

Reproduced from:

F.B. Fischl, H.H. Horowitz and T.S. Tutwiler, in "The Role of Engine Oil Viscosity in Low Temperature Cranking and Starting", Progress in Technology Vol. 10. Society of Automotive Engineers, New York, N.Y. (1966) p.110.

best with viscosity measured at $100^{o}C$ and 2.10^5 s^{-1}. Their third result was that fuel consumption in normal operating conditions correlated with viscosity measured at 100^oC and 10^6 s^{-1}. In other words, there is no question that a high-shear rate measurement would be preferable to the existing low-shear measurement, but it will be necessary to consider a number of factors when deciding on the particular rate of shear and temperature to be used in a specification. It almost goes without saying that the concept of tests at a number of temperatures and rates of shear as opposed to a single test will be accepted with the greatest reluctance by all concerned.

Whereas at high temperatures shear-thinning gives an increase in wear, which must be guarded against in formulating a motor oil, at low temperatures shear-thinning gives a benefit in easier starting. However, the full story is a little more complicated than this. It has already been stated that base oils are dewaxed in the refining process to remove paraffinic compounds which would otherwise crystallise out at low temperatures. Obviously, the depth of dewaxing is set as a compromise between the cost of dewaxing and the needs of the market. Almost invariably oils are used at temperatures below which waxes separate despite the fact that these oils will have had some of the wax fraction removed. When wax begins to separate, at a temperature characterised by the Cloud Point*, the oil becomes cloudy but it still flows readily. At a lower temperature, characterised by the Pour Point**, the oil acquires a yield stress by virtue of a network structure formed by the wax crystals. It has been estimated that a rigid structure can form in an oil with as little as 1% of solid wax. Small additions of chemicals that interfere with the formation of a structure reduce the temperature at which solidification occurs: these are known as pour-point depressants. Nevertheless, even with pour-point depressant, oils will still be used below the pour point. On the other hand, the structures thus formed can be destroyed in moderate-to-high shear and this fact is turned to good use in the cold-starting of engines: the power delivered by a starter motor is sufficient to break down wax structures at a number of degrees lower temperature than the pour point, although a limit is ultimately reached. Once the engine starts it warms up and low-temperature properties lose relevance, but before warm-up is completed it is necessary, for good lubrication, for oil to be distributed through the engine by the oil-pump, and problems arise when the oil is gelled by wax. For instance, flow to the pump inlet from the reservoir of oil in the engine sump will be impeded. In extreme cases of gelling the pump cavitates, that is, air breaks through the gelled mass and reaches the pump inlet in preference to oil. An additional problem area is the filter screen which impedes the flow of gelled oil at the entrance to the pump. Whether or not oil in the sump can flow under

* Cloud Point, ISO Method 3015, IP Method 219, ASTM Method D2500.
** Pour Point, ISO Method 3016, IP Method 15, ASTM Method D97.

its own head plus the pressure differential caused by the pump suction is mainly a function of the yield stress of the gel. Whether or not the oil can be made to flow through the screen and the oil-way system fast enough to lubricate critical components is a function of low rate of shear viscosity.

In starting an engine two major factors are involved [5] and both are dependent on viscosity. The first factor is cranking speed which must be high enough to enable a cylinder to fire; the second factor is the power that the engine delivers when it does fire and the assistance of the starter motor is disengaged. If the power is insufficient to keep the engine running it will stop. Experiment has shown that minimum cranking speed corresponds to rates of shear in the range 2000 to 8000 s^{-1}. Minimum power generation is a process that requires rather higher rates of shear. Figure 3 summarises the situation with regard to cold-starting and oil flow. It illustrates what SAE consider to be the current best thinking on the viscosities and rates of shear relevant to low-temperature engine operation and the levels of viscosity below which there should be no problem.

The above paragraphs on the rheometry of engine oils refer to gasoline and diesel automotive engine lubricants. For all other classes of oils, for example, marine-diesel oils, gas-turbine lubricants, gear oils, automatic-transmission fluids, hydraulic fluids, gas-compressor oils and machine-tool tableway oils, the standardised rheometrical methods are very much simpler. The low-shear viscosity measurement at the standard temperatures of 40°C and 100°C and the calculation of VI are always carried out. For the oils that are required to operate over a very wide range of temperatures a low-temperature, low-shear viscosity is also measured. Such oils are gas-turbine, gear and automatic-transmission lubricants. The measurement is designed to ensure that the oils will flow under modest stresses, but the stresses or shear rates are not specified explicitly.

Aviation gas-turbine lubricants are tested at various temperatures, -40°C and -53.9°C are recommended values,

using glass capillary viscometers*. Gear oils and
automatic-transmission fluids are tested with the
Brookfield viscometer** at various recommended temperatures down to -40°C. The Society of Automotive
Engineers have classified gear oils in terms of viscosity at 100°C and the (low) temperature at which
viscosity reaches 150 000 cP (see Table 3). Hence it
is now possible to have multigrade gear oils. However,
the grading scheme is quite different from that for
engine oils (Table 1) and should never be compared
with it.

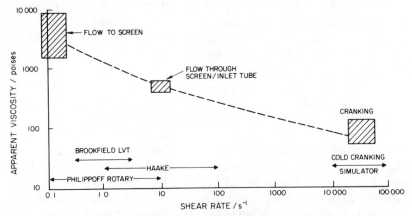

FIG. 3 – Estimated maximum oil viscosities at low temperatures

(Reproduced from 'SAE Handbook',
Society of Automotive Engineers
Warrendale, Pa.15096 (1979)
SAE Recommended Practice J 300d)

 * Aviation turbine lubricants, ASTM Method D2532.
** Brookfield viscosity, IP Method 267,
                        ASTM Method D2983.

Table 3

Axle and manual transmission
lubricant viscosity classification[c]

SAE viscosity grade	Maximum temperature for viscosity of 150 000 cP[a] $^{\circ}$C	Viscosity at 100°C[b] cSt	
		Minimum	Maximum
75W	-40	4.1	-
80W	-26	7.0	-
85W	-12	11.0	-
90	-	13.5	<24.0
140	-	24.0	<41.0
250	-	41.0	-

[a] Centipoise (cP) is the customary absolute viscosity unit and is numerically equal to the corresponding SI unit of millipascal-second (mPas).

[b] Centistokes (cSt) is the customary kinematic viscosity unit and is numerically equal to the corresponding SI unit of square millimeter per second (mm^2/s).

[c] The new viscosity classification represents a conversion to international SI units using degrees Celsius and with a minimum change in viscosity limits relative to prior practice. By early 1982, it is the aim to define the low temperature requirements at suitable multiples of 5°C while retaining 100°C for the high temperature range. The proposed revision will necessitate considering changes of the viscosity limits for the high and/or low temperatures used to define the new system.

Reproduced from "SAE Handbook"

Society of Automotive Engineers, Warrendale, Pa. 15096, (1979)

SAE Recommended Practice J306c

2.2.2 Background to research methods for low-pressure rheodynamic conditions

With the increasing use of polymers in engine and gear oils the question has often been asked whether or not viscoelastic effects have any real significance in lubrication. Over the years there have been reports of reduced wear with polymer-containing oils as compared with equiviscous (at low-shear rate) straight oils. Certainly there have been no reports of disastrously increased wear with polymer-containing oils. On the other hand the evidence of such careful work as that reported by Bell and Voisey [4] cannot be ignored. The conclusion to be drawn from this work is that the principal viscoelastic effect of significance to lubrication is shear-thinning. Although normal stresses or extensional properties were not measured there was no need to consider that any factors other than the viscosity measured at a suitable rate of shear were of major importance.

Bringing steady-state rheodynamic theory into the discussion seems only to confirm the above conclusion. Tanner [6] considered a suitable approximation to the stress equations of motion for the slider bearing and concluded that if there is to be a significant effect of normal stresses in carrying load then the ratio of the first normal stress difference, ν_1, to the shear stress, τ, must be of the order L/h where L is the bearing length and h is the minimum separation. Now conventionally L/h is a high number, and 10^3 is a not-unreasonable value. Hence the argument says that the normal stresses will influence friction and carry extra load if they are high enough, and at least 10^2 times higher than the shear stresses. The theory of slider and journal bearings was investigated in more detail by Davies and Walters [7] for the second-order and third-order simple fluid models of viscoelasticity, and an increased load-carrying effect was predicted. However, the requirement of a very large ratio of normal stress to shear stress still applied. Hence the argument rests on whether or not real lubricants in practical situations have such high degrees of elasticity as represented by this ratio.

Clearly, any further advance in this area depends on the availability of reliable measurements of first normal stress difference for typical lubricants. Some measurements have been made with rheogoniometers, but they are generally restricted to shear rates below about 10^3 s^{-1}. Furthermore, with highly-elastic liquids, secondary flows and fractures limit the upper usable level of shear rate in a rheogoniometer. In the face of these difficulties efforts have been made to use the jet-thrust technique as a means of reaching the higher rates of shear which are more relevant to steady engine-operating conditions [8]. However, the interpretation of jet-thrust measurements in terms of normal stress differences rests on the assumption of viscometric flow at the exit plane. For Newtonian and inelastic non-Newtonian liquids the departures from viscometric flow that do occur can be dealt with reasonably well by an empirical correction procedure [9]. But with polymer solutions the apparent values of normal stress difference after correction are still higher than values obtained from other measurements. Hence it must be concluded that the actual velocity and stress rearrangements at the exit plane are much more severe than those for Newtonian and inelastic shear-thinning liquids. It is believed [9] that the extra effect is a consequence of the extensional component of the flow adjacent to the wall at the exit plane, where elements of liquid slightly upstream of the exit plane, and moving very slowly, are rapidly accelerated to the much higher jet surface velocities when they move outside the tube. Many polymer solutions can exhibit abnormally high extensional stresses at high strain-rates (see Chapter 1).

An experiment has been carried out with a formulated multigrade motor oil using the torsional balance rheometer developed at University College of Wales, Aberystwyth [10]. At a maximum rate of shear of 1.48 10^4 s^{-1} the ratio v_1/τ is not greater than 0.36 at room temperature. Hence, there is little likelihood that it will reach 10^2 at 10^6 s^{-1} at the elevated temperatures typical of steady engine operation. Nevertheless there may be some formulations and some operating conditions for which v_1/τ is in the order of 10^2 or more.

In recent years increasing attention has been given to squeeze-film flows. The flow between normally-approaching surfaces is part of the complete Reynolds equation and is important in the lubrication of thrust bearings, of dynamically-loaded journal bearings and of bearings of all sorts when the relative sliding motion ceases. The characteristic feature of squeeze-film flow in the present context of elasticoviscous liquids is the extensional flow component. Theoretical modelling of this flow for elasticoviscous liquids lags behind experiment in that it predicts a decreased load-carrying capacity compared with a Newtonian liquid whereas some experiments show an increase. In this work increased load-carrying capacity means an increase in time to reach a given fractional separation between approaching parallel plane surfaces. Leider and Bird [11], using polymer solutions and a silicone fluid, found an increase in load-carrying capacity at sufficiently high loads, i.e. at rapid rates of approach. Walters and co-workers [12] showed the same effect for a solution of polyisobutene in decalin. Tichy and Winer [13] confirmed this type of result for aqueous solutions of polyacrylamide, but for oil-based polymer solutions typical of multigrade motor oils, and a silicone fluid, they found decreases. Again, as with the steady-state bearing flow problem discussed above, viscoelastic effects can influence squeeze-film lubrication beneficially, but it is unconfirmed whether they play a major part in the majority of practical situations.

An attempt has been made to measure extensional flow properties of multigrade motor oils by Oliver and Ashton using their techniques based on flow through convergent, contoured nozzles [14]. From jet-thrust measurements on the nozzle assembly, they calculated axial stress and derived axial strain from the flow rate. The fact that stress and strain were found to be proportional showed the existence of elastic behaviour. An interesting result observed in later work [15] is that whereas the shear stresses (viscosities) decrease by a factor of 7 or 8 between 22°C and 84°C, the axial elastic stresses decrease by the much smaller factor of about 2. This implies an increased relative

importance of elasticity over viscosity with increase
in temperature with those oils.

2.2.3 Background to research methods for elastohydrodynamic conditions

It is perhaps a strange fact that there are no
standardised methods available for determining rheological properties of relevance to elastohydrodynamic
(ehd) lubrication other than the zero-shear viscosity
at atmospheric pressure. Hence all the rheometry
associated with elastohydrodynamics, bar simple glass-
capillary viscometry, is confined to research
laboratories.

When two lubricated counterformal surfaces in contact move through the contact area under the influence
of a load, W, acting normal to that area (Figure 4)
the contact pressure enters the range 0.1% to 1% of
the elasticity modulus of the metal. Typical pressures
are 0.2 to 3 GPa. Lubricant is drawn into the contact
at a rate determined by the entrainment velocity,
$U = U_1 + U_2$, where it is rapidly compressed. The viscosity rises considerably above the atmospheric pressure value, $\eta(0)$, to an extent given by the following
approximate relation,

$$\eta(P) = \eta(0) \exp [\alpha P] . \tag{1}$$

Examples of viscosity-pressure relationships are
given in Figure 5 for mineral base oils and a diester
synthetic lubricant [16]. It was not possible to
extend measurements for the paraffinic oil to beyond
the pressure marked with a vertical arrow owing to
solidification of the oil. Oils of low paraffinicity
generally do not freeze under pressure. The pressure
coefficient α for the diester is about 7 $(GPa)^{-1}$
whereas values of α for mineral oils lie in the range
12 to 35 $(GPa)^{-1}$. Hence these oils reach a viscosity
10^9 times the atmospheric pressure value at pressures
of 1.73 and 0.59 GPa respectively. It is the high
viscosity which serves to separate the surfaces.

138

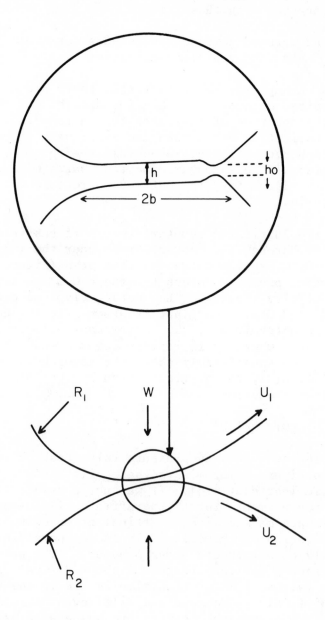

FIG.4 — Schematic of conditions in an elastohydrodynamic
contact

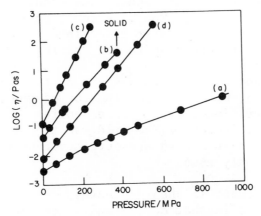

FIG. 5—Variation of viscosity with pressure:
(a) Di (2-ethylhexyl) sebacate
(b) Paraffinic mineral oil
(c) Naphthenic mineral oil at 100°F
(d) Naphthenic mineral oil at 210°F

(Reproduced from ASME 'Pressure-Viscosity Report'.
Vols. I & II 1953)

The main determining factor for satisfactory ehd lubrication is the minimum oil film thickness which, for discs in edge contact, is given by [17]

$$\frac{h_0}{R} = \frac{1.6(\alpha E)^{0.6} \ (\eta(0) \ U/ER)^{0.7}}{(W/LER)^{0.13}} , \qquad (2)$$

where R is a reduced radius of curvature ($1/R = 1/R_1 + 1/R_2$), L is the contact length perpendicular to the direction of motion and E is a reduced Young's modulus.*

It is noteworthy that h_0 is only slightly dependent on load, in contrast to classical hydrodynamics in which $\eta(0) \ ULR/W$ appears as a relevant group, and that

* In a revision of the film-thickness formula, Dowson writes the first two factors in the numerator as $2.65(\alpha E)^{0.54}$ and the other factors remain unchanged.

the viscosity enters as the atmospheric pressure value under the temperature and shear rate conditions existing in the inlet zone. The dependence on Young's modulus is even less than it is on load, so the lubricant film is not sensitive to the bearing material.

The important material parameters of the lubricant are $\eta(0)$ and α, hence a determination of the pressure coefficient of viscosity, α, is essential in predicting ehd film thickness. The Bridgman technique of a falling plummet in a high pressure environment is the one most commonly used.

If the lubricant is shear-thinning, the value of $\eta(0)$ to be used is that determined at the mean rate of shear in the inlet region [18]. Strictly, the value of α should also be determined at that rate of shear. In the little work that has been done to determine viscosity as a function of rate of shear at high pressures, capillary and rotational concentric cylinder viscometers have been used.

In addition to the minimum lubricant film thickness the other important factor in ehd lubrication is the frictional force or, as engineers prefer, the traction. A knowledge of the coefficient of friction is of importance in determining the losses in a lubricated contact, but an additional incentive in recent years to know more about the friction in ehd contacts has been the increasing potentiality of the traction drive as a fuel-efficient, variable-speed transmission. Continuously variable traction drives employ metal surfaces rolling together in nominal line or point contact with a lubricant between. Traction is transmitted from one member to the other through the lubricant film which is operating in elastohydrodynamic conditions.

In elastohydrodynamics research, great use is made of the twin-disc machine in which discs in edge contact (nominal line contact) are rotated at preset peripheral velocities, U_1 and U_2, under an applied load W (Figure 4). The pressure distribution and the contact area are, in practice, close to the results

given by the classical elasticity analysis of Hertz.
Apart from the slight constriction at the exit the
surfaces in the contact are parallel and the ratio of
the contact width, 2b, to the central film thickness,
h, is sufficiently large that simple shear conditions
can be considered to exist.

Experiments with the disc machine show that the
friction observed in constant load experiments, in
which the rolling speed U is maintained constant,
increases with sliding speed $\Delta U = U_1 - U_2$ up to a
maximum value and then decreases (Figure 6). Furthermore, there is an upper limiting curve for a given
fluid (the dashed line in Figure 6) and values of
coefficient of friction above this limit are not
accessible whatever the applied load or the rolling
speed. It is now generally considered that the fall
in friction at high sliding speeds beyond the maximum
friction is predominantly an effect of viscous self-
heating. The rest of the results arise from the very
complex rheological behaviour of the lubricant.

At low levels of viscosity, or at high temperatures
and low pressures, the initial linear rise of friction
with ΔU reflects Newtonian viscous behaviour, which
changes to shear-thinning viscous behaviour through
the peak in the friction. Self-heating is insufficient
to account for all the shear-thinning observed. At
high levels of viscosity and at moderate-to-high
pressures in the ehd range, many lubricants change
from a liquid to a glass in their passage through the
contact and the initial linear rise of friction with
ΔU reflects the behaviour of an elastic solid. This
response changes to plastic flow at higher sliding
speeds up to the friction peak. An additional factor,
which will have to be taken into account in a complete
theory, is that of volume retardation. This is the
lag behind the instant of application of a step
function in pressure for the volume of the fluid, and
hence the viscosity and shear modulus, to reach
equilibrium values. Volume retardation, sometimes
referred to in the literature as structural relaxation,
is of lessening significance to the occurrence of the
liquid-to-glass transition the higher the contact
pressure.

142

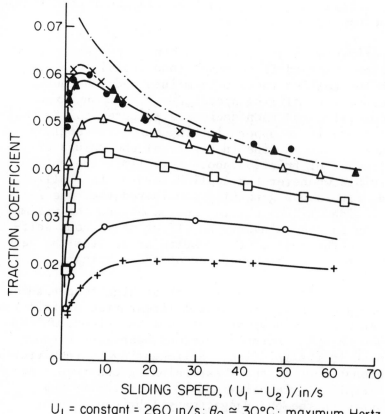

U_1 = constant = 260 in/s; $\theta_0 \simeq$ 30°C; maximum Hertz pressure in 10^3 lbf/in^2 : + 62, o 87, □ 110, △ 147, ▲ 176, ● 200, × 224.

FIG. 6 – Variation of traction with sliding
speed at various contact pressures

(Reproduced from K.L. Johnson and R. Cameron
Proc. Inst. Mech. Eng. (1967) 182(1) 307)

Various models of rheological behaviour have been put forward to explain the friction results. The most successful is that of Johnson and Tevaarwerk [19] which is of a non-linear Maxwell model with shear modulus G and non-linear viscosity given by a hyperbolic sine function as follows:

$$\frac{1}{G} \dot{p}_{ik} + \frac{p_{ik}}{p_e} F(p_e) = 2e_{ik}^{(1)} , \qquad (3)$$

where

$$p_e = [(p_{ik} p_{ik})/2]^{\frac{1}{2}} , \qquad (4)$$

$$F(p_e) = \frac{p_o}{\eta} \sinh(p_e/p_o) , \qquad (5)$$

where p_o is an adjustable parameter which the authors call a representative stress.

At the highest pressures, those of greatest relevance to traction drives, the model of an elastic solid with a yield stress (the elastic-plastic model) is adequate. In this model there is no viscous term.

Figure 7 (taken from Johnson and Tevaarwerk [19]) shows theoretical curves and experimental points for the frictional force as a function of rate of strain, both expressed non-dimensionally, for various values of the Deborah number of the disc-machine experiment. The Deborah number is $\eta U/Gb$, the ratio of the relaxation time of the lubricant at the mean Hertzian pressure involved (η/G) to the mean transit time through the contact (b/U). The diagram illustrates the main features discussed above, namely the linear viscous initial region followed by a non-linear viscous region at low Deborah numbers, the linear elastic region followed by plastic flow at high Deborah numbers, and the close approximation to elastic-plastic behaviour at the highest Deborah numbers.

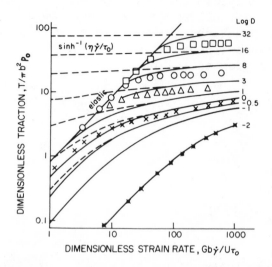

FIG. 7 – Theoretical non-dimensional traction curves for varying Deborah
number D(= $\eta U/Gb$). Experimental results for four lubricants
at a mean pressure of 0.67 GPa and one lubricant at 0.47 GPa
(*Reproduced from K.L. Johnson and J.L. Tevaarwerk, Proc. Roy. Soc.,
(1977) A356 215*)

In an approximation for large rates of strain,
Hirst and Moore [20] derive the following expression
for the friction coefficient.

$$\mu = \alpha p_o - \frac{p_o}{p} \ln \left\{ \frac{p_o}{2\eta(0)\dot{\gamma}} \right\} , \qquad (6)$$

where $\dot{\gamma}$ is the steady simple shear rate. This
predicts a limiting value

$$\mu_{max} = \alpha p_o ,$$

which is a result obtained earlier by B.K. Daniels.

For the complete characterisation of lubricants for
friction properties in ehd lubrication we need to know
the three material functions

$$\left.\begin{array}{l} \eta = \eta(P,T) \ , \\[4pt] G = G(P,T) \ , \\[4pt] p_o = p_o(P,T) \ . \end{array}\right\} \tag{7}$$

But in the approximations two parameters are needed: in Hirst and Moore's approximation, η and p_o, and in the elastic-plastic approximation, G and yield stress, p_c, are needed as functions of pressure and temperature.

2.3 Some rheometrical methods for oils

2.3.1 Low shear, atmospheric pressure methods

The standard methods ISO 3104, IP 71, ASTM D.445 and ASTM D.2532 all use glass-capillary viscometers and gravity flow to measure kinematic viscosity, ν. Dynamic viscosity, η, is then obtained, if required, from the density, ρ, determined at the same temperature, and the formula,

$$\eta = \nu\rho \ . \tag{8}$$

Kinematic viscosity is calculated, from the flow time, t, required for the specified volume of oil to flow, from the equation,

$$\nu = Ct \ , \tag{9}$$

where C is a known calibration constant. Limits are placed on the flow time in order that the additional terms of a more exact form of the above equation, and ascribed to kinetic energy effects, can be neglected. Precision between laboratories (reproducibility) is 0.7% for transparent oils at temperatures between 15° and 100^oC.

Calibration depends on the absolute value for pure water at 20.00^oC which is taken as 1.0020 mPas, or 1.0038 cSt. This value is uncertain to $\pm0.25\%$. Calibration of working viscometers is best carried out in authorised centres by comparison with master

viscometers which have, in their turn, a calibration
derived from the water value. However, it is permiss-
ible to use standardised oils which are available from
various suppliers whose addresses can be obtained on
application to such bodies as the Institute of
Petroleum.

It cannot be emphasised too strongly that the essence
of precise viscosity measurement is precise control and
accurate measurement of the temperature of the determination, since it is not unusual for the viscosity of
an oil to change by as much as 10% for every degree
Celsius change in temperature.

For dark, opaque oils the so-called reverse-flow
capillary viscometer is used. In this type of viscometer the oil flows upwards past the timing marks,
rather than downwards as in the "direct" flow instrument, in order that the position of the meniscus should
not be obscured by oil left behind on the glass wall.
It is generally accepted that the accuracy and precision of the reverse-flow method is a little worse
than that of the "direct" flow method.

Outside the USA suspended level viscometers of the
Ubbelohde or BS/IP/SL (British Stands/Institute of
Petroleum) type are the most common "direct" flow
instruments (Figure 8). They have a slight operational
advantage over the Cannon Fenske, which is favoured in
the USA, since they need no correction for temperature
when the test temperature differs from the calibration
temperature. Of the reverse-flow instruments the
BS/IP/RF (Figure 9) and the Cannon Fenske Opaque are
widely used. Automatic viscometers based on gravity
flow through a glass capillary are available for
laboratories whose work load can justify the cost.

The measurement of viscosity at low temperatures with
the Brookfield viscometer by the methods IP/267 and
ASTM D.2983 for transmission and gear oils is carried
out with the LVT form of the instrument and spindle 4.
Three samples of the oil are contained in glass tubes.
After the samples have reached the test temperature,
one is tested at increasing spindle speeds (eight are

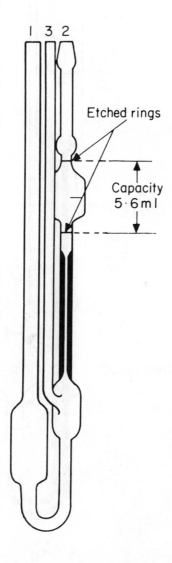

FIG. 8 – BS/ IP/SL viscometer

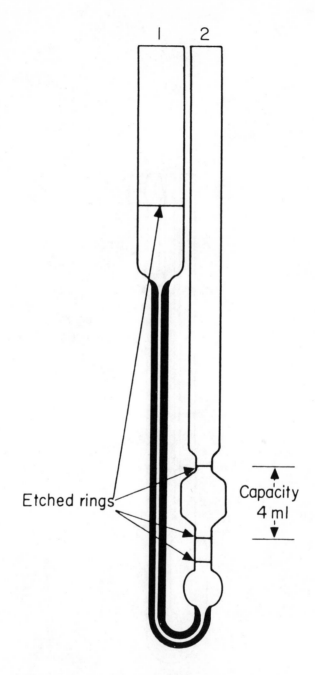

FIG. 9 – BS/IP/RF U-tube
reverse-flow viscometer

available from 0.3 to 60 rev/min) until the instrument
scale reading is between the centre and the upper limit
of the scale. The speed at which this happens is then
the test speed for the other samples which then provide
duplicate results. The scale reading is converted to
viscosity using a factor obtained by calibration with
selected Newtonian oils. Hence the measurement is
carried out with a concentric cylinder viscometer with
outer radius 11.25 mm and inner radius 1.58 mm. It is
effectively a measurement at constant shear stress
distribution, the maximum value being about 34 Pa at
the inner cylinder wall. The range of maximum rate of
shear from lowest to highest speed, for Newtonian
liquids, is 0.016 to 3.2 s^{-1}. The reproducibility
depends on the type of oil and the temperature and
varies from 12% to 50%. Co-operative studies are in
hand to improve the precision which is probably a
function mainly of temperature measurement and control.

The Brookfield viscometer is one of the candidate
instruments for a proposed low rate of shear, low-
temperature measurement on motor oils. Others under
consideration in Europe include the Haake Rotovisko
RV2, which is a constant speed, concentric cylinder
instrument, the Hoeppler rolling ball viscometer and
a U-tube capillary viscometer operated according to a
modification of the German standard DIN 51568. In the
USA a similar range of instruments has been under
investigation, but views are converging on the Cannon
Mini-Rotary viscometer, which is a Stormer type, that
is, a constant shear stress, concentric cylinder
instrument driven by falling weights.

The Mini-Rotary can be adapted to give the yield
stress measurement as well as viscosity and this
instrument is a strong contender for the yield stress
measurement in the proposed new SAE specification for
motor oils. On both sides of the Atlantic, the Vacuum-
Pipette apparatus has also been used for the yield
stress measurement. In this instrument the pressure
drop required to start a column of oil moving is determined.

2.3.2 High shear, atmospheric pressure methods

Two classes of instrument are used: capillary tubes
and concentric cylinders.

A capillary instrument widely used in Germany is the
Umstätter Structurviskosimeter [21]. The oil sample
is driven by a constant applied air pressure from a
compressed-air source through the capillary and the
flow is measured in a burette attached to the upper
end of the vertical capillary. An empirical procedure
based on Newtonian oils of known viscosity is applied
to correct for self-heating which becomes noticeable
between 5.10^4 s^{-1} and the upper limit of 10^5 s^{-1} rate
of shear.

Wright and Mather [22] have described a new concentric cylinder viscometer for use with polymer-containing
oils. It is shown in Figure 10. High rates of shear
of the order 10^6 s^{-1} are achieved by the use of annular
gaps of about 2 μm. The final drive to the inner rotor
is a steel spindle, 3 mm diameter and 280 mm long,
equipped with universal joints, and this allows the
alignment of rotor and stator. Shear heat is removed
by circulating thermostat fluid inside the rotor as
well as outside the stator. Torque is measured on the
stator from the deflection of suspension springs. A
unique feature of the design is the measurement of the
annular gap whilst a test is in progress. This is
achieved by measuring the electrical capacitance between rotor and stator and making allowance for the
permittivity of the oil. Although the capacitance
method measures an average gap width, it is the same
average as that required for the determination of viscosity from the total torque on the stator, hence the
electrical method is ideally suited to the purpose.
It was found with Newtonian oils that no correction
was necessary for self-heating at rates of shear up to
10^5 s^{-1}. Beyond 10^5 s^{-1} empirical methods are used
for the correction since it was found that theoretical
methods underestimated the temperature rise. The
viscosity of Newtonian oils is measurable to within
±5%.

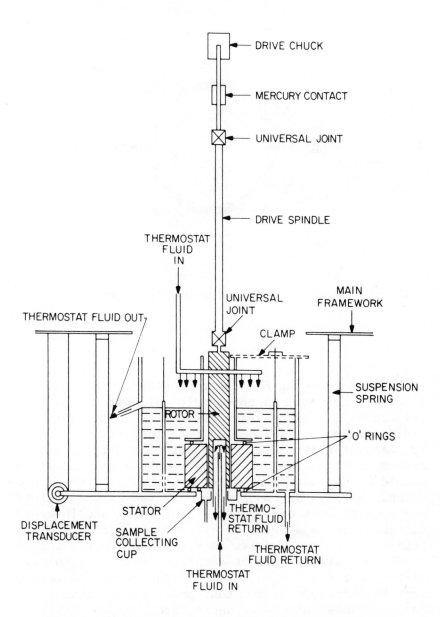

FIG. 10 – High shear rate concentric cylinder viscometer

(Reproduced from B. Wright and J. Mather,
SAE technical paper 790 212 (1979)

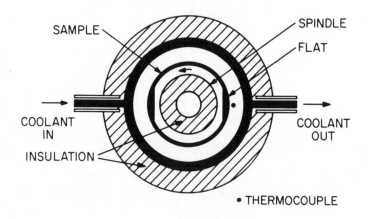

FIG.11— Cold Cranking Simulator viscometric cell (schematic)

The Cold-Cranking Simulator of standard method ASTM
D.2602 is basically a concentric cylinder viscometer
(Figure 11). The bath temperature is controlled to
-18°C and sample is injected into the viscometer space
by means of a syringe. The instrument differs from
conventional rheometers in two respects and therefore
should be considered as a test rig rather than a rheometer. First, the inner rotor has two flats machined
at opposite ends of a diameter and these are intended
to introduce an element of hydrodynamic wedge action.
Secondly, the rotor drive is a universal motor which
can be operated either at constant input a.c. voltage
or constant input d.c. current. The resulting measurement is therefore at constant power rather than at
constant shear stress or shear rate. Reproducibility
of results is ±12%.

2.3.3 High pressure methods

The most widely used technique for measuring viscosity at low rate of shear at high pressures is the Bridgman falling-body technique. The viscosity is given by the formula

$$\eta = k(\rho_1 - \rho_2)t \, , \tag{8}$$

where ρ_1 and ρ_2 are the densities of the plummet and oil respectively, k is a calibration constant and t is the time to fall a known distance vertically through the oil. It is usual to determine k by reference to known oils. For pressures up to about 0.3 GPa the transducers used to detect the position of the plummet, and hence determine t, can be positioned outside the high-pressure environment. Above this pressure the transducers must be mounted inside and this makes for an appreciable increase in complexity of the apparatus. With manual timing the technique is limited to the viscosity range 10^{-3} to 10^3 Pas with an accuracy better than 5%. An automatic instrument capable of reaching 1.4 GPa and 10^4 Pas has been described in some detail by Irving and Barlow [23].

For the measurement of viscosity at high rates of shear and high pressures capillary instruments have been used mainly, although Hamilton [24] has constructed and successfully used a concentric cylinder apparatus. The capillary instrument of Novak and Winer [25] is shown in Figure 12. The test fluid is contained in the reservoirs R_1 and R_2 and the interconnecting tubing. It is separated from hydraulic fluid contained in spaces I to IV. By pressurising space I and venting II, the test fluid is brought to the required high pressure up to a limit of 0.7 GPa. By adjusting the difference in pressure between spaces III and IV, the translating piston is made to drive the test fluid through the capillary. Using capillaries of 0.01 inch nominal bore and various lengths, the shear stress range 30 Pa to 5 MPa has been covered. Kinetic energy and self-heating effects have been considered by Jakobsen and Winer [26]. For another capillary viscometer, Galvin, Hutton and Jones [27] have observed for

154

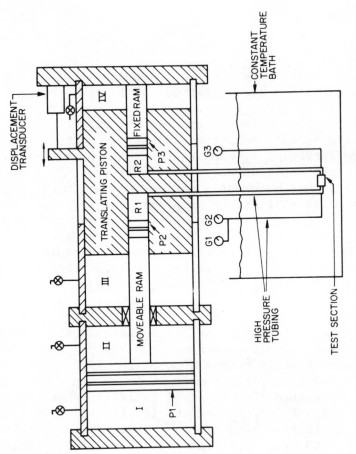

FIG.I2— Schematic drawing of high pressure capillary viscometer
(Reproduced from J.D. Novak and W.O. Winer, Trans.
ASME (F) J. Lub. Tech. (1968) 90 580)

base oils that the self-heating correction of Galili, Takserman-Krozer and Rigbi [28] is adequate, that is, within 5%, over a range of conditions. In their apparatus, the isothermal correction was found to be more appropriate than the adiabatic correction. With this correction, the upper limit of rate of shear can be extended by a factor of about 3. They also found that by making an empirical adjustment to the correction factor the effective rate of shear range could be extended still further, to rates of shear about 5 times that at which self-heating was first observed. With light-to-medium viscosity lubricating oils this upper rate of shear is about 3.10^5 s^{-1} and the upper limit of pressure so far studied is 0.2 GPa.

High-frequency oscillatory measurements on lubricating oils and related liquids were pioneered by Lamb and his co-workers [29]. Hutton and co-workers [30] have also made these types of measurement with particular reference to high ambient pressure conditions. The principal use of such measurement in modern elastohydrodynamics is to provide values of the shear modulus, G, under the conditions of temperature and pressure that exist in the contact. A second use is to provide information about the time delay from the instant of application of the high contact pressure to the attainment of the corresponding, smaller, equilibrium specific volume (volume retardation). A further use, which has to some extent been superseded by other methods, is to provide an estimate of the non-Newtonian properties of the lubricant in ehd conditions by use of the well-known analogy between the viscosity-rate of shear function and the viscosity-frequency function.

The techniques are mainly based on the measurement of the change in amplitude and phase of a pulsed train of mechanical waves on reflection at the interface between fused quartz and the oil. Most of the work has been done in the range 6-78 MHz frequency, using waves generated piezoelectrically. The reflected waves are also detected piezoelectrically.

With suitably-cut piezoelectric crystals, experiments can be carried out with either shear waves or ultrasonic waves. Because ultrasonic waves can penetrate liquids much farther than shear waves, it is possible to use wave-propagation techniques as well as wave-reflection techniques for determining mechanical properties. The additional measurements by wave-propagation are of velocity and attenuation.

The equipment and techniques involved are complicated and too specialised to describe here. The reader should consult the books by Harrison [31] and Matheson [32].

It is customary to express the amplitude change and phase change in terms of the real part, R, and the imaginary part, X, of a complex mechanical impedance, Z, defined by,

$$Z = R + iX . \tag{9}$$

Then the complex shear modulus (analogous expressions apply for ultrasonic (longitudinal) waves) is given by,

$$\left.\begin{aligned} G* &= G' + iG'' , \\ G' &= (R^2 - X^2)/\rho \simeq R^2/\rho , \\ G'' &= 2RX/\rho . \end{aligned}\right\} \tag{10}$$

It is often difficult to determine X, and at high pressures X has not been determined at all, hence the results are often expressed in terms of the impedance component R. Over much of the stress relaxation range R >> X. Of course, at frequencies where the modulus G for ehd calculations is measured, X = 0.

For the ultrasonic experiments, the velocity and attenuation are expressed in terms of a complex longitudinal modulus which can be written,

$$M* = M' + iM'' . \tag{11}$$

The complex bulk modulus, K*, is related to the measured quantities by

$$M* = K* + (4/3)G* . \tag{12}$$

Hence, by carrying out both shear and ultrasonic experiments, the shear and bulk properties can be determined.

For making measurements at high pressures the pressure vessel has a working volume about 25 mm in diameter and 240 mm in length. All apparatus for high-frequency experiments and for the necessary subsidiary measurements of low-shear viscosity and density is designed to operate in this space. Provision is made for separating the test oil from the pressurising fluid. Electrical leads are taken out through static seals, whilst for the operation of an acoustic absorptiometer and a rotating concentric cylinder viscometer a specially designed seal allows a rotating shaft to be brought into the high-pressure region. Although this shaft seal can hold a pressure of 0.7 GPa it allows a rotary motion with very little resistance.

The high-frequency rheological properties of oils can also be determined by optical techniques [33,34]. The measurement depends on the existence in the liquid of Debye elastic waves caused by the thermal motions of the molecules. These waves propagate in all directions and are acoustic in nature. Hence they are capable, in principle, of giving information about both shear and bulk deformation properties. If a linearly polarised laser beam is passed through the liquid, the scattered light is in two parts. The first part is scattered from non-propagating density fluctuations and the second part is scattered from the Debye waves. The intensity of the scattering is determined by the mean spacing of the scattering centres, e.g. the wavelength of the Debye waves, and by the wavelength of the light according to the Bragg condition for scattering from a lattice. Since the Debye waves are propagating, there is a Doppler shift of the frequency of the light probe, whilst the light scattered from the non-propagating fluctuations is not frequency shifted. Hence the scattered light, analysed with a Fabry-Perot interferometer, is detected as three peaks (Figure 13), with a central Rayleigh line at the reference frequency $\omega_o = 0$ and the Doppler-shifted Brillouin lines. The Doppler-shift of frequency, ω_B, enables the velocity of the Debye acoustic waves to be calculated by

$$V_L = \frac{\omega_B \lambda}{2n \sin (\theta/2)} , \qquad (13)$$

where λ is the wavelength of the incident light, θ is the scattering angle and n is the refractive index of the liquid. The longitudinal modulus component M' is then

$$M' = \rho(V_L)^2 . \qquad (14)$$

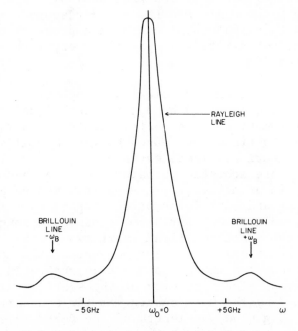

FIG. 13 — Frequency spectrum from a Brillouin scattering experiment

If depolarised light is used, the resulting three-line spectrum is associated with shearing motions, hence the determination of the Brillouin shift in this case leads to a measurement of the real component of the shear modulus,

$$G' = \rho(V_S)^2 . \qquad (15)$$

This measurement requires the oil to be in a highly-viscous condition, near the glassy state.

In principle it is possible also to determine relaxation functions, but little work has been published so far. When volume, or shear, relaxation occurs this means that the liquid structure cannot respond in an equilibrium way to the acoustic wave motion. Hence scattering intensity is transferred from the Brillouin lines to the Rayleigh line. The enhanced Rayleigh line is now called a Mountain line. Digital correlation spectroscopy is required to get the relaxation function from the data.

Since the frequency of the acoustic waves examined by the Brillouin scattering techniques is in the GHz range, it is not necessary to confine experiments to supercooling liquids, as is required with the piezoelectric methods. The main strength of the method is in determining the high-frequency limit of the shear modulus, the G used in elastohydrodynamics, and in detecting the onset of the glassy state. In addition, a relatively simple high-pressure cell, requiring only two windows, can be used.

Efforts are now being made to measure rheological properties of oils at high pressures under large-strain conditions. Three pieces of apparatus for this purpose have been described by Scott Bair and Winer [35]. Two of these apparatuses are for measuring shear stress/strain curves at pressures up to 0.7 GPa and 1.2 GPa respectively. The former is shown in Figure 14. Shearing is similar for both and takes place in an annulus of fluid on a central cylindrical surface. The axial motion is controlled by the pressure drop across the central driving piston. Again an isolating piston is used to separate the sample from the pressurising fluid. In the glassy state the shear modulus and yield stress of the oil are measured.

The other apparatus is a high shear stress viscometer based on simple shear flow between a fixed cylindrical surface and a concentric rod which is

160

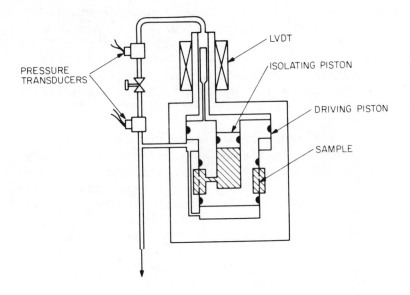

Fig 14.-Schematic of high pressure shear stress apparatus
*(Reproduced from Scott Bair and W. O. Winer, Trans.
ASME (F) J. Lubric. Tech. (1979). 101 251)*

driven axially (Figure 15). Movement of the rod is
detected electrically with a transducer and the driving force is calculated from the pressure drop across
the driving piston. A passageway to the side of the
shearing gap (not shown) allows fluid displaced by
the central rod to flow in the opposite direction and
thereby maintain the pre-set ambient pressure.

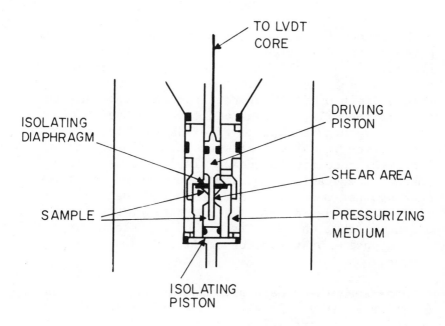

FIG. 15 — Schematic of high shear stress viscometer
(*Reproduced from Scott Bair and W. O. Winer,
Trans. ASME (F) J. Lubric. Tech. (1979)*
101 251

3. RHEOMETRY OF LUBRICATING GREASES

3.1 Nature of lubricating greases

The common feature of all greases is that they comprise a liquid lubricant and a thickening agent. Mineral oils are the most widely-used liquid components and metal soaps of one sort or another are used as the thickener in over 90% of total production. As with lubricating oils, various additives may be incorporated to improve resistance to oxidation, to improve lubrication in extreme pressure conditions, and so on. Additives specific to greases include those which improve the stability of the structure of the thickener against mechanical and/or thermal degradation. Since most greases have a yield stress, it is not essential that the additives should dissolve or stay in suspension by virtue of small size and Brownian motion.

Metal-soap thickeners are made by saponification of a fat or a fatty acid with the appropriate alkali. Sometimes the saponification is carried out in the grease plant in the presence of some of the oil, whilst in other processes pre-formed soaps are brought in. Following saponification, or the dissolving of soap at a high temperature, the remainder of the grease-making process is essentially to crystallise the soap in an optimum way. The optimum may be selected on the basis of economics - the soap is more expensive than the oil - or on some performance criterion.

Views on the structure of grease have not changed in recent years. It has been shown [36] that the basic soap crystal is ribbon-shaped with variable ratios of dimensions. However, ratios of 500:10:1 are not untypical. Aggregates of crystalline material also form: these have a less-elongated shape. The ribbons and aggregates interact to form a three-dimensional structure which traps the oil by capillary forces and gives the mass a yield stress.

Cheap greases for applications not far removed from room temperature are based on calcium soaps. High

quality greases, which melt at about 180°C and operate
in high-speed rolling bearings quite well at 120°C,
are based on lithium hydroxystearate soap. Higher
operating temperatures can be attained by the so-
called complex greases which are made with either two
soaps, one having a short and the other a long hydrocarbon group, or combinations of soaps and salts, e.g.
borates. The highest melting points are achieved with
non-soap greases of which the commonest are thickened
with finely-divided clay. The shape of the basic
thickener particle in the clay grease is plate-like,
i.e. one small dimension and two almost equal large
dimensions. In recent years, the polyureas have been
used as thickeners in greases that require special
qualities, e.g. oxidation resistance, in high-
temperature operation.

3.2 Why rheometry?

3.2.1 Background to standardised methods

In the oil lubrication of all but small and lightly
loaded devices, such as watches and scientific instruments, it is necessary to provide a reservoir of oil,
an oil-feed arrangement and, often, a filter. These
provisions can be bulky, expensive and require skilled
maintenance. By virtue of its yield stress, a grease
stays put in its housing and is available for lubrication and, if properly designed, enables the engineer
to circumvent the complications associated with the
use of oil in certain cases. The rolling bearing, for
example, properly lubricated with the correct grease
is a compact, relatively cheap and efficient component
which requires little in the way of maintenance. Indeed, the lubricated-for-life rolling-element bearing
is used in significant numbers, and this bearing is
maintenance free.

The yield stress of a grease is not measured on a
routing basis; however all greases are tested for
consistency or, in the jargon of the technology,
penetration*. The consistency is measured by dropping

*Penetration of greases, IP Methods 50 and 310,
ASTM Methods D217 and D1403.

a specified cone from a given height into a sample
prepared in a specified way. After a sinking time of
5 s the cone is stopped and the depth of penetration
is measured in units of 10^{-4} m. Greases are graded
according to the penetration determined at 25°C, as
shown in Table 4. As an example, a grade 2 grease has
a penetration between 265 and 295 units.

Table 4

N.L.G.I. consistency classification of greases

Grade Number	Worked penetration range 10^{-4} m at 25°C
000	445 – 475
00	400 – 430
0	355 – 385
1	310 – 340
2	265 – 295
3	220 – 250
4	175 – 205
5	130 – 160
6	85 – 115

It should be obvious that the variation of yield
stress, or consistency, with temperature is of some
importance since greases are used over wide ranges of
temperature. In a recently developed test procedure*,
an apparatus called the trident-probe and based on a
Brookfield viscometer is used to measure consistency
continuously whilst the grease is heated at a uniform
rate. In rheological terms, the measure of consistency
involved is rather obscure since the shearing device
is a set of three prongs that is rotated whilst immersed in the grease contained in an annular space. Sometimes a measurement cannot be made because the probe
carves out a channel in the grease and the torque drops
to very low meaningless values.

* Flow properties at high temperatures, ASTM Method
    D3232.

The other standardised grease rheometrical method is for the measurement of apparent viscosity*. The end use of such measurements is principally for the calculation of pressure drops or flow rates in the pipes of centralised lubrication systems. Two classes of such systems are the grease dispenser of the automotive service station and the industrial installation that feeds the bearings of a plant at regulated intervals from a central point. It is estimated that in the automotive application the rates of shear involved range from 10 to 7000 s^{-1}, whilst in the industrial system the range is 0.1 to 100 s^{-1}.

Figure 16 shows experimental results obtained with capillary viscometers. The approach of the curve to a straight line of slope -1 at low rates of shear indicates the existence of a yield stress [37]. At high rates of shear the curve approaches, but does not reach, the line of slope zero which indicates the viscosity of the base oil from which the grease is made, at the temperature of measurement. The results show the effects of soap content and base oil viscosity. Soap content has the greatest effect at low rates of shear, compare curves A and D. Base oil viscosity has the greatest effect at high rates of shear, compare curves A, B and C.

3.2.2 Background to research methods

Greases are used in many situations but their commonest and often most-demanding use is in rolling-element bearings, that is, ball- and roller-bearings. The state of lubrication in these bearings is primarily elastohydrodynamic. Now, no work has been done to determine the rheological properties of greases in ehd conditions and the reason for this is twofold: first, because of the experimental difficulties involved and secondly, because it has been thought that a grease at high rates of shear has a viscosity similar to that of the mineral oil from which it is made, plus a small thickening effect of soap particles according to a

* Apparent viscosity, ASTM Method D1092.

166

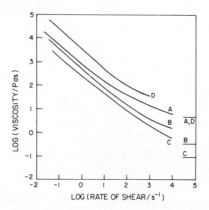

FIG.16—Capillary viscometer results for lithium soap-
      based greases. The corresponding base-oil
      viscosity is indicated on the right hand side.

A. 8 % soap −20°C
B. 8 % soap   0°C
C. 8 % soap 20°C
D. 14 % soap −20°C

(Reproduced from D. Klamann, L. Endom, R. Rost &
A. Haak. Erdöl und Kohle (1967) 20 219)

variant of the Einstein equation for non—interacting
suspensions. Associated with this idea is the belief
that the pressure coefficient of viscosity at high
shear rate is the same as that of the base oil. There
is a great deal of truth in these ideas, and the fact
that a grease structure seems ultimately to be almost
completely and permanently destroyed in the shear
conditions of an ehd contact supports the view that
the lubricant is a suspension of virtually non-
interacting soap particles.

However, experiment shows otherwise. The elastohydrodynamic film thickness of a grease—lubricated
contact, from experiments done with roller—bearings as
well as disc machines, has been shown [38] to start at
a value well above that for the base oil and then, with
continued running, to fall to a value below that for
the base oil determined at the same inlet temperature.
Typically, the final film thickness in the absence of
replenishment of the lubricant is 70% of the base oil

value. Hence there is a substantial discrepancy which
is thought to be due to starvation in the inlet to the
contact. The starvation, or lack of lubricant, must
originate in the rheological properties of the grease
and is worthy of investigation.

Lest it should be thought that a grease is less
satisfactory than its base oil in respect of film
thickness, the description given in the previous para-
graph does not tell the whole story. Owing to the
starvation effect less energy is dissipated in the
inlet to the ehd contact so the heating of the grease
is less than that of the base oil under fully-flooded
inlet conditions. Hence the loss of film thickness
by starvation is nearly compensated by the gain of
film thickness arising from the higher viscosity
corresponding to the cooler inlet.

So much for the elastohydrodynamic contact. However,
there are other important features of grease lubrication. When a ball-bearing is packed with grease the
operator tries to fill all the accessible spaces.
Consequently when the bearing is started up the moving
parts are forced to move through the grease and,
except for the small ehd contact, this is a low rate
of shear process. Since the viscosity is very high
there is considerable resistance to the motion and
heat is generated. Considerable heat is also generated in the elastohydrodynamic films. The friction and
heat are both undesirable features, particularly with
high-speed bearings. It is therefore necessary for
the grease to be so designed that it moves out of the
path of the balls, but not so much that no lubricant
is left behind. The displaced grease then moves into
the recesses of the bearing end-plates, also known as
covers or shields. It is also desirable that the
displaced grease should not flow back too rapidly for
the churning to be re-established. On the other hand
the grease is expected to lubricate for long periods
of time, and completely static grease in the covers
cannot take part in lubrication, apart from a process
of separation of oil from the structure, known as
bleeding. Finally, it is known from experience with
good greases that the clearances formed between the

static grease and the moving parts are very small.
These provide effective seals against dirt and con-
tamination from outside and this sealing effect is a
real bonus from a good grease.

The whole process of starting up a new bearing and
the attainment of steady running conditions in a few
hours or days is called clearing. It can be seen
that clearing is a complex of different, sometimes
competing, effects amongst which a balance has to be
reached. At one extreme the clearing must not be so
ineffective that the grease churns and the bearing
runs hot, nor must it be so effective that the grease
is lost from the balls or, in extreme cases, lost
from the covers as well, a process known as winding-
out.

A number of rheological parameters are relevant to
clearing. First, the rate of generation of heat is a
function of the viscosity, which itself depends on
the base oil viscosity and the yield stress. The
lower the rate of shear involved the greater is the
influence of yield stress over that of base oil.
Secondly, once a channel has been cut by the balls
the flow back is determined by the yield stress.
Thirdly, the movement of grease from high rate of
shear regions to low rate of shear regions can be
influenced by viscoelasticity. It has been shown [39]
that the elasticity that is revealed by the existence
of first normal stress differences has a big influence
on clearing. A carefully adjusted degree of elasticity will provide the required optimum clearability.
However, if the first normal stress difference is too
high the grease winds out of the bearing completely.

Thus the grease formulator has three rheological
parameters, and their variation with temperature and
rate of shear, at hand to design a grease.

A great variety of instruments has been used to
determine the rheological properties and none is free
from difficulties. All instruments will have a wall
effect since the structure of a grease near a wall
will be different from that in the bulk. This effect

is greater the smaller the shearing gap of the instrument. In capillary viscometers a length/diameter ratio of the capillary of at least 40 is required to avoid end-effects. In cone-and-plate and parallel-plate rheometers fracturing of the grease may occur to give spurious results and a similar effect has also been observed in concentric cylinder instruments. Finally, in the measurement of normal stresses the yield stress of the grease makes it impossible to determine the zero of normal force with conventional rheogoniometers. For this measurement the yield stress problem is minimised with the torsional balance rheometer.

3.3 Some rheometrical methods for greases

The penetration cone of the IP 50, ASTM D217 tests is double-angled: the main part has an included angle of 90° and the sharper tip an angle of 30°. The length of the tip is 15 mm, so for grade 6 greases the penetration is insufficient to cause more than the tip to be immersed in the sample. At the other end of the consistency range, with the grade 00 and 000 greases, the cone is in danger of being flooded so a special one is used. The so-called semi-fluid greases include grade 000 materials and some that are softer still. Some of them are pourable. Sometimes a falling-cylinder viscometer is used in controlling the consistency of these materials.

Grease penetration is measured at 25^oC on an unworked and a worked sample. The worker is a cylindrical metal pot with a movable perforated circular plate which is pumped up and down through a grease sample for a standard 60 strokes. The aim is to bring the grease to a reproducible mechanical condition for the consistency test, since on standing the recovery of the yield structure can be a long process and a measurement made on an unworked grease will then depend very much on its storage history.

Most greases are made batchwise and all batches are controlled on penetration. However, since this procedure is time-consuming some plants use a Ferranti-

Shirley cone-and-plate viscometer to control production, having first established a correlation between
the viscometer measurement and the penetration for the
particular grease. The Ferranti-Shirley allows a
measurement to be made in a few minutes as compared
with half-an-hour or more for the temperature-
stabilisation of a penetration test. Generally a low
rate of shear of 10 to 30 s^{-1} is used, since a viscosity
in this range is known to correlate with penetration.

Attempts have been made to replace penetration with
a yield stress measurement in which smaller stresses
and lower deformation rates are applied. None has
successfully replaced penetration as a routine basis.
In one, the Cone Resistance Value test (CRV) [36], a
right-angled cone was allowed to sink for a minute -
or more - to allow it to more-nearly approach its
equilibrium penetration depth. The CRV was calculated
as the net downward force divided by the projected
area of the cone at the grease surface.

As discussed in an earlier review [36], there is no
correct yet practical way of determining the yield
stress of a grease since the flow curve has no sudden
transition between flow and no-flow. The test of a
good method is whether the results obtained from it
enable one to predict an aspect of grease performance
in service conditions.

In the standard viscosity method, ASTM D1092, grease
at the selected temperature is forced through a
capillary at a selected flow rate. The pressure drop
required is measured with gauges. Apparent viscosity
is then calculated from the Poiseuille equation for a
Newtonian fluid. The arrangement of the equipment is
shown in Figure 17. Grease in the reservoir, S, is
separated from hydraulic oil by a piston, L. A con-
stant-delivery pump, A, generates the flow. Eight
capillaries, C, each of 40:1 length:diameter ratio,
and two pump speeds enable measurements to be taken at
sixteen different shear rates. The lowest mean
apparent rate of shear available is 10 s^{-1}, which is
two orders of magnitude higher than the range applicable to large-scale centralised lubrication systems.

In some laboratories larger capillaries are used to extend the range. The Shell-de Limon viscometer was designed for an extended range of shear rates. It is basically like the ASTM instrument, but the capillary is many times longer and coiled up to fit into a thermostat.

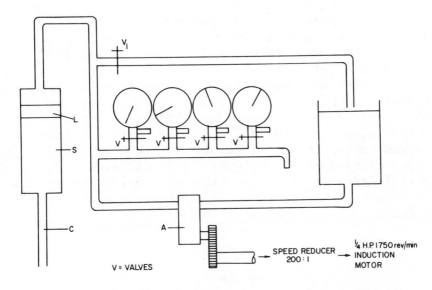

FIG. 17 — Arrangement of the grease viscometer of method ASTMD 1092

Cone-and-plate and parallel-plate rheometers have been used. To avoid slip at the wall it is recommended [40] that the plates be grooved or roughened. It is necessary to space the grooves close together so that the major part of the shearing boundary is made up by grease held in the grooves.

If shear is continued at a constant rate for a period of time at stresses in the vicinity of the yield stress a fracture surface will generally appear in the grease in the cone-plate space [41]. The torque then falls and spurious stresses are obtained unless suitable correction is made for the decreased shearing radius. However, it is usually possible, by applying much higher rotational speeds, to pass out of this condition

172

and attain a state in which the grease behaves like a liquid. Whether it then continues to show a fracture instability depends on the first normal stress difference, which for many greases is rather low.

It is not possible to measure the first normal stress difference with a grease using a Weissenberg, or similar, rheogoniometer. The reason is that the zero of the normal force measurement in this instrument is determined at zero rotational speed and the zero of the normal force spring cannot be attained with a material that has a yield stress. The reason for that is that the yield stress prevents the required radial flow between the platens. On the other hand, when the sheared grease is behaving as a liquid, radial flow can occur and the normal force deflection corresponding to this condition is the true one. Hence it is possible only to obtain differences in normal force at different rates of shear. The problem of normal stress measurement is overcome by using the torsional balance rheometer [42].

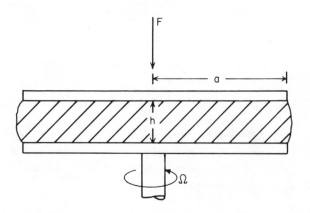

FIG. 18 — Schematic diagram of the platen arrangement
in the torsional balance rheometer
(Reproduced from D. M. Binding, J. F. Hutton
and K. Walters, Rheol. Acta, (1976) 15 540

In this instrument a known normal force, F, is applied to parallel plates, radii a, of which one is free to move along the axis joining the centres (Figure 18). The rotation, Ω, is applied, and the final equilibrium separation of the plates, h, marks the equilibrium of this applied force with the normal force generated by the grease at the rate of shear $a\Omega/h$. The lower limit of rate of shear achieved with this rheometer is that at which a sufficient part of the specimen in the centre of the plates cannot be considered to be in the liquid condition. No more-exact specification of this limit has been made. A useful procedure with the torsional balance rheometer is to apply a method of bracketting. In this method a rotational speed is first selected at which the pre-set platen separation increases towards equilibrium; Obviously the required equilibrium lies between these two pre-set separations.

174

REFERENCES : CHAPTER 3

[1]   C.M. Murphy, J.B. Romans and W.A. Zisman,
       Trans. ASME 71 (1949) 561

[2]   W. Philippoff, ASLE Trans. 1 (1958) 82

[3]   F.B. Fischl, H.H. Horowitz and T.S. Tutwiler, in
       "The Role of Engine Oil Viscosity in Low
       Temperature Cranking and Starting", Progress
       in Technology Vol. 10. Society of Automotive
       Engineers, New York, N.Y., 1966, p.110

[4]   J.C. Bell and M.A. Voisey, in "The Relationship
       between Engine Oil Viscosity and Engine
       Performance", Vol. 1. Society of Automotive
       Engineers Special Publication, SAE SP 416,
       New York, N.Y., 1977, p.71

[5]   G.K. Vick, in "The Role of Engine Oil Viscosity
       in Low Temperature Cranking and Starting",
       Progress in Technology, Vol. 10. Society of
       Automotive Engineers, New York, N.Y., 1966,
       p.160

[6]   R.I. Tanner, ASLE Trans. 10 (1967) 313

[7]   J.M. Davies and K. Walters, in "The Rheology of
       Lubricants" ed. T.C. Davenport, Applied
       Science Publishers Ltd., Barking, 1973, p.65

[8]   J.M. Davies, J.F. Hutton and K. Walters, in
       "Polymeres et Lubrification", ed. C. Wolff,
       Colloques Int. du C.N.R.S. No. 233, Centre
       Nationale de la Recherche Scientifique, Paris,
       1975, p.61

[9]   J.M. Davies, J.F. Hutton and K. Walters, J. Non-
       Newtonian Fluid Mech. 3 (1977/78) 141

[10]   D.M. Binding and K. Walters, J. Non-Newtonian
       Fluid Mech. 1 (1976) 277

[11]   P.J. Leider and R.B. Bird, Ind. Eng. Chem.
       Fundamentals 13 (1974) 342

[12] G. Brindley, J.M. Davies and K. Walters, J. Non-
      Newtonian Fluid Mech. $\underline{1}$ (1976) 19

[13] J.A. Tichy and W.O. Winer, Trans. ASME Series F;
      J. Lubric. Technol. $\underline{100}$ (1978) 56

[14] D.R. Oliver and R.C. Ashton, J. Non-Newtonian
      Fluid Mech. $\underline{2}$ (1977) 367

[15] D.R. Oliver and R.C. Ashton, J. Non-Newtonian
      Fluid Mech. $\underline{4}$ (1979) 345

[16] American Society of Mechanical Engineers,
      "Pressure-Viscosity Report" Vols. I and II,
      1953

[17] D. Dowson and G.R. Higginson, "Elastohydrodynamic
      Lubrication - The Fundamentals of Roller and
      Gear Lubrication", Pergamon Press, London,
      1966

[18] A. Dyson and A.R. Wilson, Proc. Inst. Mech. Eng.
      $\underline{180}$ Part 3K (1965-66) 97

[19] K.L. Johnson and J.L. Tevaarwerk, Proc. Roy.
      Soc. $\underline{A356}$ (1977) 215

[20] W. Hirst and A.J. Moore, Proc. Roy. Soc. $\underline{A365}$
      (1979) 537

[21] K. Kirschke, Rheol. Acta $\underline{7}$ (1968) 354

[22] B. Wright and J. Mather, SAE Technical Paper
      790212 (1979)

[23] J.B. Irving and A.J. Barlow, J. Phys. E. Sci.
      Instr. $\underline{4}$ (1971) 232

[24] G.M. Hamilton, Reading University, unpublished.

[25] J.D. Novak and W.O. Winer, Trans. ASME(F) J. Lub.
      Tech. $\underline{90}$ (1968) 580

[26] J. Jakobsen and W.O. Winer, Trans. ASME(F) J.
      Lub. Tech. $\underline{97}$ (1975) 472

[27] G.D. Galvin, J.F. Hutton and B. Jones, Paper presented to British Society of Rheology meeting, Aberystwyth, September 1979

[28] N. Galili, R. Takserman-Krozer and Z. Rigbi, Rheol. Acta 14 (1975) 550

[29] J. Lamb, J. Rheology 22 (1978) 37, for a recent review.

[30] J.F. Hutton, M.C. Phillips, Jessie Ellis, G. Powell and E. Wyn Jones, "Elastohydrodynamics and Related Topics" ed. C.M. Taylor, M. Godet and D. Berthe, Mechanical Engineering Publications Ltd., Bury St. Edmunds, 1979, p.173, for a recent summary.

[31] G. Harrison, "The Dynamic Properties of Supercooled Liquids", Academic Press, London, 1976

[32] A.J. Matheson, "Molecular Acoustics", Wiley, London, 1971

[33] J.F. Dill, P.W. Drake and T.A. Litovitz, ASLE Transactions 18 (1975) 202

[34] M.A. Alsaad, W.O. Winer, F.D. Medina and D.C. O'Shea, Trans. ASME(F) J. Lub. Tech. 100 (1978) 418

[35] Scott Bair and W.O. Winer, Trans. ASME(F) J. Lub. Tech. 101 (1979) 251

[36] J.F. Hutton, in "Principles of Lubrication", ed. A. Cameron, Longmans, London, 1966, p.521, for a review.

[37] D. Klamann, L. Endom, R. Rost and A. Haak, Erdöl und Kohle 20 (1967) 219

[38] A.R. Wilson, Proc. Inst. Mech. Eng. 193 (1979) 185

[39] J.F. Hutton, in "Proceedings of the JSLE-ASLE International Lubrication Conference, Tokyo, 1975", ed. T. Sakurai, Elsevier, Amsterdam, 1976 , p.707

[40] G.R. Dobson and Anne Tompsett, in "The Rheology of Lubricants", ed. T.C. Davenport, Applied Science Publishers, Barking, 1973 , p.96

[41] J.F. Hutton, in "The Rheology of Lubricants", ed. T.C. Davenport, Applied Science Publishers, Barking, 1973 , p.108

[42] D.M. Binding, J.F. Hutton and K. Walters, Rheol. Acta 15 (1976) 540

CHAPTER 4

FOODS

By

M. Van den Tempel

CONTENTS

v. few specific examples

	Page

J

1. THE APPLICATION OF RHEOMETRY

Food technology suffers from tradition; many industrial
processes are still based on operations similar to
those carried out by housewives or craftsmen several
centuries ago. Subjective assessment of quality is
widely used, both for raw materials and for the final
product. Attempts to measure consistency under conditions that are relevant to those employed by the consumer have resulted in an impressive variety of instruments and test methods. Such imitative tests are useful for control purposes, but they do not generally
provide physically meaningful parameters. Extensive
reviews [1,2] are available, and we need only refer to
these methods in so far as they can be used to obtain
information about rheological parameters.

The application of rheometry in the food industries
is mainly concerned with the control of the consistency
of the final product, and not so much with the behaviour of materials in processing operations. One
reason is that during processing, the consistency (or
even the composition) is usually modified quite drastically, by aeration, mixing, homogenization, crystalli-
zation by cooling, etc. The fluid mechanics of materials that change their properties during flow has not
yet reached the stage where it can be applied to the
complicated flow fields encountered in the food industries. This is unfortunate, since a detailed

understanding of the relations between processing conditions and product consistency would be helpful in
making products with controlled consistency. In cases
where legal or economic constraints prohibit a change
in composition this is often the only available route
for modifying consistency. Even in big, modern industries the use of this route is still largely based on
trial and error instead of on a judicious use of
proper rheological measurements.

Meanwhile, however, product consistency is still the
main area of application of food rheology. The ultimate judge of consistency is the consumer, when handling the food in the kitchen, on the plate, or when
eating it. Therefore, the ultimate reason for most of
the work in food rheology is to satisfy the demands of
consumers. They expect certain attributes in food
products, with respect to consistency or texture. It
is interesting to be able to measure these attributes,
and such measurements can often be used in factory
control. However, in using the results of rheometric
experiments, one tries to go one step further; one
aims at being able to make products of the desired,
controlled consistency. This requires relations to be
established between, on the one hand, the subjective
consumers' assessment and, on the other hand, the composition and manufacturing process. It is then convenient to express the results of panel (or consumers')
assessment in terms of physically meaningful, i.e.
rheological parameters. These, in turn, can be related to an "internal structure" of the material, which
is fully determined by composition and process (including storage conditions). Fig. 1 illustrates the
various relationships. In this chapter, we shall be
mainly concerned with the measurement of rheological
parameters, and their relation to subjective assessment and internal structure.

2. SUBJECTIVE ASSESSMENT

The relation between the behaviour in actual use
(which may include behaviour in a manufacturing process) and rheological parameters can, in general, not
be established by a careful analysis of the deformations and stresses to which the material is subjected

182

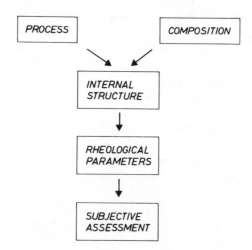

Fig.1 Showing how rheological parameters relate subjective assessment to the manufacturing process.

in use. The materials of interest are usually inhomogeneous on a microscopic or colloidal scale, their constitutive equation is in most cases unknown and it is expected to be complicated. The deformation pattern is far too complicated, and in many cases the material changes during actual use as a result of heating, mixing with saliva, etc. No great help can be expected from the use of "imitative" tests, except in the simplest possible cases where such tests are not really necessary.

Under these conditions, the obvious way to establish the desired relations is by the statistical analysis of correlations. In using a hedonic scale for subjective attributes of a set of similar materials, and at the same time subjecting these materials to a series of properly selected rheological tests, it is normally found that consumers can perceive variations in certain well-defined rheological parameters. It is also found that the number of mutually independent rheological properties that determine the perceptible consistency is much smaller than the number of possible tests. The test results, both subjective and rheological, will then consist of various combinations of these few properties: some attributes depend on 2 or 3 independent properties, and some may even show a perfect correlation with one single property. If it is assumed that the combinations are linear, Factor Analysis [3] can be used to determine the contribution of each of these properties to each of the test results. In particular, one can then identify test methods that respond to only one particular property. Recent applications of this technique to fish products and to textured vegetable protein [4,5] seem to suggest that the consistency of meat-like products is determined by no more than two independent rheological properties: tenderness and juiciness. Although the latter property is not generally classified as a rheological one, it appears to correlate well with measurements of the Poisson ratio measured in unilateral compression [6].

In the following, it will be assumed that relations have been established between certain rheological tests and results of subjective measurements. The rheological parameters must then be used to derive structural information.

3. INTERNAL STRUCTURE OF FOODS

The internal structure of the material is meant to describe the way in which the constituents are distributed in space, and the way in which they are held together to produce perceptible elasticity, or plasticity. The constituents may be macromolecules, particles, emulsion droplets, air bubbles, or combinations thereof.

The emphasis is on "perceptible", which means that we shall only be interested in the structural elements that contribute to elasticity in a time scale corresponding to the time of observation. For consumers, this is roughly between 10^{-1} and say, 10^3 sec. Many processing operations have a characteristic time of about 10^{-1} to 1 sec. Hence, isolated molecules do not contribute to perceptible elasticity or plasticity, since even macromolecules have a longest relaxation time shorter than about 10^{-2} sec. Similarly, orientation or deformation of single units (particles, droplets, or molecules) does not usually account for elastic properties in the required time range. Thus, deformation of an emulsion droplet of radius R and interfacial tension σ in a flow field [7] occurs in a time of the order of $\eta R/\sigma$, which is less than 10^{-2} sec for most cases of interest. Longer times require a decreased mobility of the constituents, obtained by their coherence in larger structural elements: aggregates, or a network. These are the structural elements that must be considered in discussing the visco-elastic or non-Newtonian properties of foods.

Two problems arise in the interpretation of rheological data in terms of a model structure. The first problem is concerned with the starting point: does one believe that the material properties are mainly determined by a polymer network, or should one consider the material as a dispersion of interacting particles? In the former case, one can fall back on the vast amount of knowledge about polymer solutions that has been collected over several decades [8]. Unfortunately, though most foods do contain macromolecules in the form of proteins and/or polysaccharides, their conformation and behaviour is usually quite different from that of the simple model systems studied by polymer rheologists. In particular, the time-temperature superposition principle cannot be used for foods: a change in temperature would alter the composition of the material by melting of fat or ice crystals, changing the conformation of protein molecules, gelling, etc. Moreover, the rheological behaviour of foods becomes non-linear even at relatively small deformations.

Most foods contain dispersed particles or emulsion droplets of a colloidal (< 1 μm) or microscopic size, and it has been shown that colloidal interactions between such particles may give rise to visco-elastic and non-Newtonian properties somewhat similar to those encountered in polymer solutions [9,10]. The colloidal interactions may even be due to polymer adsorbed at the particle surface [11].

It is obvious that it would be helpful if a model

structure could be selected that does not depend on *a priori* knowledge of the behaviour of certain constituents, and that can be used in systems containing both particles and polymer. Such a model will be proposed in section 6.

The second problem is concerned with the analysis of the structure by means of mechanical spectroscopy. This technique uses small deformations, so that the structure is not destroyed during and by the measurement. It results in properties such as the storage and loss modulus as a function of the frequency and of the magnitude of the deformation, and the creep compliance for characterizing the behaviour at longer times (> 5 sec). These results can in principle be translated into a relaxation spectrum, which contains the same information but in a somewhat more convenient way.

As is well-known, the results of mechanical spectroscopy are rarely sufficient for characterizing a material structure, unless the type of structure that is to be considered is already known from other investigations. Many cases have been recorded [12-15] of particulate dispersions (usually crystals of metal soap in oil) exhibiting in small or slow deformations a behaviour that cannot be distinguished from that of certain polymer solutions.

It should be noted that the results of mechanical spectroscopy do not correlate well with consumers' or panel assessment of consistency [16]. Handling by consumers generally involves the destruction of the original structure, and the consumer wants the destruction to proceed in a manner which is typical of a given product. Therefore, consumers' assessment is well-correlated with results of experiments using large deformations, in particular with the ultimate properties of the material.

More information about structural details may be obtained by using techniques for measuring in the region where the behaviour starts to become non-linear. These are suitable for studying the beginning of structural breakdown in dispersions. Such techniques

include superposition of steady state flow and small-
amplitude oscillations, or large amplitude oscillatory
shear [17-21]. The former method allows, in principle,
measurement of the amount of structure still left
after the application of a given rate or magnitude of
the deformation. The second technique is of interest
because non-linear behaviour is exhibited by the
appearance of higher harmonics in liquid-like but not
in solid-like systems [21]. The difference in
behaviour is undoubtedly related to the strong inhomogeneity of the deformation in solid-like systems [22].

A prerequisite for the use of these techniques is
the availability of a model theory that relates experimental results to structural parameters. The
models to be discussed in section 6 might be suitable
for this purpose, but results have not yet been published.

4. THE MEASUREMENT

The selection of proper experiments should be based
on the following requirements:
(i) the results should correlate well with certain
aspects of subjective assessment, and
(ii) the results should be expressed in rheologically
meaningful parameters.
The first requirement suggests the use of transient,
large deformations, because consumers are not generally interested in steady state flow properties. The
second requirement restricts the possibilities to
simple geometry: simple shear or elongation. Compression (squeezing flow) is still difficult to describe
in terms of well-defined rheological properties [23],
in particular for materials exhibiting complicated
rheological behaviour.

Let us consider how the structure of a plastic
material is mechanically destroyed under geometrically
well-defined conditions. We start with the undisturbed material, at the origin of the coordinate system
of Fig. 2. Under the influence of a gradually increasing stress, the material is deformed more and
more rapidly. As soon as the stress reaches a critical value, the structure is broken down so far that

further deformation becomes much more rapid even at
low stress values. The behaviour becomes more liquid-
like, instead of the original solid-like behaviour.
If now the stress is gradually decreased we may follow
a different line BC; the difference is called thixotropy. The down-line can be extrapolated to the ab-
scissa to give a "Bingham yield value". It is important to realize that this yield value relates to the
rate of structural build-up after the structure had
been more or less destroyed. It does not give information about the strength of the original structure.
The strength (point A) is often called yield value,
and this is the property that correlates well with the
results of penetrometer or compression tests. It is
often widely different from the Bingham yield value.

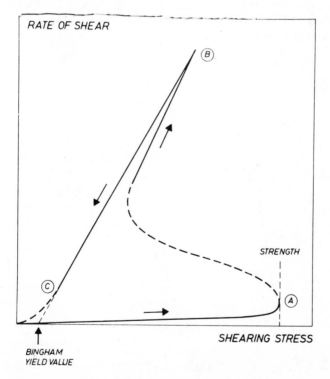

Fig.2 Schematic representation of structural
    breakdown in a visco-elastic and plastic
    material subjected to a gradually
    increasing shearing stress.

The literature contains several examples of measurements corresponding to Fig. 2 [24]. Nevertheless, the technique is complicated, partly because the interrupted part of the curve represents unstable behaviour, and cannot be measured. The same information can, however, be obtained more conveniently from a measurement of stress overshoot in a start-shear experiment.

The main difference with the test of Fig. 2 is that we now apply a constant rate of deformation, starting at $t = 0$, and measure the shearing stress and the first normal stress difference required to maintain that strain rate. Low strain rates are preferably used to avoid inertia effects. Fig. 3 shows typical results for a simple shearing deformation. In certain materials it may be more convenient, or more relevant, to use elongational flow, but it is still unsettled whether stress overshoot will occur in elongation [25].

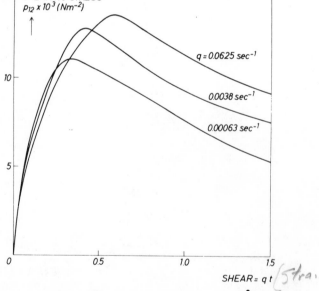

Fig.3 Stress-strain curves for butter at 10°C in simple shear.

Again starting with the undisturbed material, the stress initially increases more or less proportionally to the deformation. This might suggest that we are dealing with an ideal elastic material, and that the slope of the straight line would represent an elastic modulus. This, however, is not so.

A large and increasing amount of structural break-down occurs when the stress is increased along the initial straight line. This follows from two observations: (i) Upon sudden removal of the stress, recovery is incomplete. This proves that a certain amount of permanent deformation has occurred; the deformation is not entirely elastic and the slope is not an elastic modulus. (ii) The elastic modulus determined at very small deformations (e.g. $\gamma < 0.01$) is often up to several orders of magnitude larger than would correspond to the slope of the straight line. Actually, it is generally possible to measure a catastrophic drop in the storage modulus if the amplitude of the deformation exceeds 1%, or in some materials even at deformation in excess of 10^{-4} [17].

What really happens in the initial, straight part of the stress-strain curve of Fig. 2 is a competition between structural breakdown and re-formation. At the maximum stress overshoot value, re-formation can no longer compete. Further deformation would ultimately result in a horizontal straight line at a lower stress level. The ratio between the ultimate stress and the strain rate would correspond to a viscosity. It can, however, only be interpreted as a viscosity if the material deforms uniformly. In many materials, in particular inhomogeneous food products, this is not the case.

The maximum stress value in a start-shear experiment corresponds to the strength of Fig. 2. Moreover, within a group of similar materials it correlates well with the subjective impression of "hardness", or with its measurement by penetrometer or compression tests. The deformation at which the maximum stress occurs is even more interesting: it measures toughness, or longness, vs shortness or brittleness or crispness.

The process occurring at and near the maximum stress overshoot is similar to rupture, and it can be discussed in terms of the "classical" theory of rupture of visco-elastic materials [26], without invoking the use of a constitutive equation. This theory considers

190

the stress concentration near the tip of a "weak
region" in or at the surface of the material. The
high local stress relaxes through the same mechanisms
that give rise to the equilibrium relaxation spectrum.
In a given experiment, the characteristic time for
stress increase near a tip may correspond to the time
of a major relaxation mechanism; in that case one
finds that the ultimate properties (i.e. the position
of the stress maximum) depends on the strain rate.
Fig. 4 shows an example; in this case the relaxation
mechanism responsible for the strain-rate dependence
has been identified as being due to flow of continuous
liquid between and around the dispersed droplets [17].

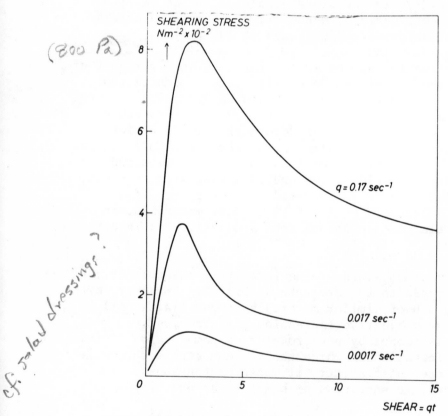

Fig. 4 Stress-strain curves of a 20%
emulsion of water in oil.

For very brittle materials, like biscuit or a choco-
late bar, the stress drops rapidly to zero from its
maximum value, which is situated at a very small
strain. Measurement of the strain at break is, of
course, essential for the assessment of brittleness,
about which no information can be obtained from the
measurement of breaking stress only.

5. STRUCTURAL IMPLICATIONS

It is important to realize that stress-overshoot in
start-shear is a common property of non-linear viscoelastic materials, whether they be polymer solutions
or particle dispersions. The phenomenon appears to be
independent of the mechanism responsible for elasticity, which is mainly of an entropic nature in polymeric systems and energetic in dispersions or emulsions.
The common type of behaviour in start-shear experiments suggests that these widely differing materials
have common structural elements which are responsible
for this behaviour.

These non-linear visco-elastic materials have many
more properties in common. In small-deformation
rheometry, they show:
(a) G' decreasing with decreasing frequency;
(b) G" = $\eta'\omega \neq 0$ in plateau region; tg $\delta \simeq 0.1$;
(c) G' decreasing with increasing deformation beyond
    a small linear region.

Fig. 5 shows some examples for biopolymer gels and
for particle gels, as they occur in several types of
food products. In larger deformations the following
range of properties is usually found:
(d) Stress overshoot in start shear;— *if rate controlled.*
(e) Plasticity, permanent set;
(f) Storage hardening or thixotropy;
(g) Self-healing, up to spinnability in elongation;
(h) Syneresis if liquid-continuous;
(i) Low mechanical strength, compared to, for example,
    polymeric constructional materials.

All visco-elastic food products exhibit these properties in varying degrees, and this is what makes them
suitable as foods. The common structural element must

be such that it allows storage of elastic energy over a period of the order of 1 sec. It is likely that the main relaxation mechanism operating at around 1 sec. is the same in all these materials. It consists of breaking and re-formation of junctions in a temporary network structure.

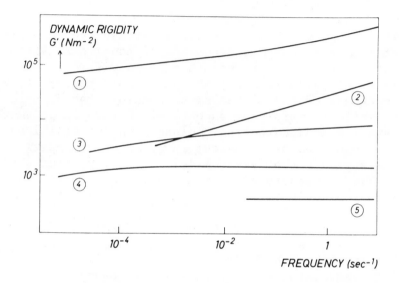

Fig.5 Storage moduli of typical food products and model systems. (1) Soy protein gel, tg δ ≃ 0·3; (2) Processed cheese, tg δ ≃ 0·25; (3) Margarine, tg δ ≃ 0·1; (4) Alginate gel, tg δ ≃ 0·01; (5) Pectin gel, tg δ ≃ 0·01.

6. TEMPORARY NETWORKS

The study of temporary networks started shortly after James and Guth [27] had definitely established the permanent network model for crosslinked rubberlike materials. Tobolsky and co-workers [28,29] considered a rubberlike polymer containing a known number of cross-links made in a state of rest. The material was

then deformed, a fraction of the crosslinks was removed and a certain number of new crosslinks added.
What are the properties of the material? The solution
of this problem gave rise to the 2-network model [30].

Tobolsky had to work with permanently crosslinked
polymers, in which the crosslink density could only be
changed by rather drastic means: high temperature or
chemical agents. Meanwhile, it had been established
that the elastic properties of (non-crosslinked)
"concentrated" polymer solutions were due to physical
entanglements that are constantly removed and reformed,
due to Brownian motion. So long as their life-time is
long compared to the characteristic time of motion of
network segments, they act as temporary crosslinks.
Deformation affects the mechanical properties of the
network by modifying the rates of removal and of
creation of entanglements.

This model was incorporated into the constitutive
equation by Lodge [31], using a logical generalization
of the 2-network theory. The result will be written
as:

$$p'_{ik} = \int_{-\infty}^{t} \frac{G(t')}{\lambda(t')} \exp\left(-\int_{t'}^{t} \frac{dt''}{\lambda(t'')}\right) S_{ik}(t')dt' . \quad (1)$$

The relaxation time λ is the probability for a junction to leave the network in unit time. G is proportional to the rate of creation of junctions; the
proportionality factor is the contribution of each
junction to the stress (i.e. kT for a rubberlike
polymer). The kernel of equation (1) represents the
number of junctions formed at time $t' < t$ and still
present at the time of observation t. The strain
tensor S_{ik} describes the deformation between t' and t.

Although most of the work on temporary networks has
been carried out with a view to entanglement junctions,
it is now clear that the same formalism can be used
for any type of weak, reversible bond between network
units. Such bonds may be due to van der Waals forces,
dipolar or other electrostatic (ionic) interactions, or
the attraction between polymer segments under conditions where the Flory-Huggins parameter approaches 0.5.

In general, there will exist a distribution of bond
strengths, and hence of their probability of rupture
or creation. This is reflected in the relaxation
spectrum (Fig. 6) as determined by mechanical spectroscopy in the region of linear behaviour [8]. The relaxation spectrum does not contain more information
than the results of measurements from which it was obtained, but its introduction is often convenient for
correlating results of various types of measurement.
Moreover, one may hope that the spectrum can be interpreted in terms of processes occurring on a molecular
or colloidal scale.

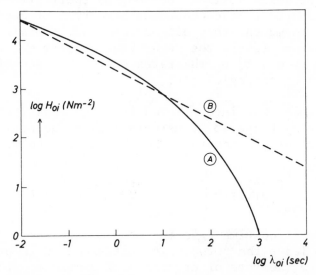

*Fig.6 Typical relaxation spectra for a polymeric
system (A) and for a plastic dispersion (B).*

The spectrum is usually a continuous one, and the
measurable properties contain integrals over the
entire spectrum. Nevertheless, in actual calculations,
a summation is substituted for the integration. The
relaxation mechanisms operating in the range between
$\log \lambda_n$ and $(\log \lambda_n + \Delta \log \lambda)$ contribute an amount $G_n =
H_n \Delta \ln \lambda$ to the modulus. Equation (1) is valid for
each of the relaxation mechanisms $n = 1, 2 \ldots \infty$, and
the total stress tensor (in so far as it is due to the
network) is simply made up of the sum of the contributions of all the mechanisms:

$$p'_{12} = \sum_n p'_{12(n)} \cdot$$ (2)

The problem is now to find how the material parameters G_n and λ_n are affected by a deformation. For
entanglements, the kinetics of their creation and
removal was introduced in equation (1) by Meister [32],
using concepts developed by Graessley [33]. Although
objections have been raised against certain aspects
of the physical model, we shall consider Meister's
theory in somewhat more detail because of its mathematical simplicity, and because it predicts all the
properties listed under (a) - (i) in section 5.

Restricting our considerations to a simple shearing
motion at a shear rate q, it is postulated that this
will reduce all relaxation times in the following way:

$$\lambda'_n = \frac{\lambda_n}{1 + cq\lambda_n} \cdot$$ (3)

For simplicity, the moduli G_n are decreased in the
same way, such that the ratio G_n/λ_n is not affected
by the deformation. The value of the constant c is
obtained from a detailed consideration of molecular
interactions; for entanglement formation in simple
shear the result is c = 0.295.

Substituting (3) into (1) and integrating gives, for
steady state shear,

$$\eta_s = \sum_n G_n \frac{\lambda_n}{(1 + cq\lambda_n)^2} ,$$ (4)

and for the shear stress overshoot in start shear:

$$\eta(t) = \sum_n \frac{G_n\lambda_n}{(1 + cq\lambda_n)^2} \left[1 - \left\{ 1 - cqt - c^2q^2t\lambda_n \right\} \exp\left(-\frac{t}{\lambda'_n} \right) \right] \cdot$$

(5)

These results are illustrated in Fig. 7 to demonstrate the fairly realistic behaviour of this theory,
which does not contain any adjustable parameter! The
maximum stress overshoot occurs at a deformation
$qt_m = 1/c$, which seems to confirm our belief that the
toughness property depends mainly on the ease at
which broken junctions are re-formed; i.e. the self-
healing capacity. It would be easy to adjust Meister's
result in order to obtain better agreement with experiment. This will not be attempted here, mainly because
there are physical objections against the use of the
rate of deformation in equation (3) [34,35]. It is
better to use the strain itself, or the stress, as the
parameter determining the shift of the relaxation
spectrum. This, however, makes the mathematics much
more complicated and closed expressions such as
equation (5) can no longer be obtained.

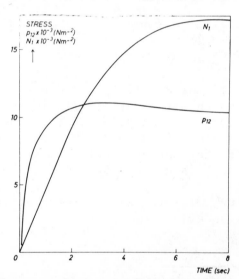

Fig.7 Stress overshoot as predicted by Meister's
theory, for relaxation spectrum A of Fig. 6
at $q = 1 \cdot 0 \, sec^{-1}$ and $c = 0 \cdot 295$

Marrucci and co-workers [36,37] use the elastic
stored energy as the parameter that governs the
effect of deformation on the relaxation mechanisms.
In order to facilitate calculations they start from
the general equation for a temporary network in
differential form:

$$\frac{p'_{ik(n)}}{G_n} + \lambda_n \frac{\delta}{\delta t} \left[\frac{p'_{ik(n)}}{G_n} \right] = 2\lambda_n e^{(1)}_{ik} \qquad (6)$$

This is a generalized Maxwell equation, and it can be shown to be fully equivalent to the integral equation (1), under certain conditions which we shall not consider here.

The generalization of the well-known one-dimensional Maxwell model (spring and dashpot in series) is obtained by:

(i) introducing the extra stress tensor p'_{ik} and the rate of strain tensor $e^{(1)}_{ik}$;

(ii) using a convected time derivative

$$\frac{\delta p_{ik}}{\delta t} = \frac{dp_{ik}}{dt} + \frac{\partial v^m}{\partial x^i} p^{mk} + \frac{\partial v^m}{\partial x^k} p^{im} ; \qquad (7)$$

(iii) summing over all the contributions of the relaxation spectrum as in equation (3), and

(iv) making the material parameters λ_n and G_n dependent on a structure factor x:

$$G_n = G_{on} x_n ; \qquad \lambda_n = \lambda_{on} x_n^{1 \cdot 4} . \qquad (8)$$

The peculiar exponent in the second equation (8) is semi-empirical, and relates particularly to the behaviour of entanglements. The structure factor x_n is the fraction of junctions of type n still present in the strained material. In the unstrained state, all $x_n = 1$ and every x_n is characterized by a λ_{on} and a G_{on} as found from the relaxation spectrum in the region of linear behaviour. Deformation decreases the value of x_n (i.e. entanglements are removed) as shown in the final term of the kinetic equation:

$$\frac{dx_n}{dt} = \frac{1 - x_n}{\lambda_n} - a \frac{x_n}{\lambda_n} \sqrt{\frac{p_{11(n)}}{2 G_n}} . \qquad (9)$$

Here, p_{11} (or rather the first normal stress difference) stands for the elastic stored energy, and a is
the only adjustable parameter in the theory. The
first term on the RHS of equation (9) represents
creation of junctions, at a rate proportional to the
concentration of junctions, that have been broken.
Comparison of predicted and experimental results shows
satisfactory agreement for several polymeric systems
under a wide range of conditions, using values of a
between 0.4 and 0.8 [38–42]. As an example, Fig. 8
shows shear stress overshoot curves as predicted by
this theory, for the two relaxation spectra of Fig. 6.

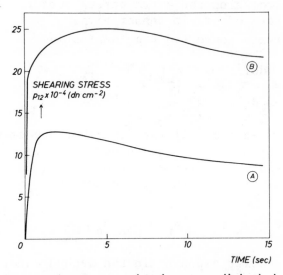

Fig.8 Shear stress overshoot as predicted by
Marrucci's theory, for relaxation spectra
of Fig.6: a = 0·4; q = 2·5 sec $^{-1}$.

The theory is attractive because it contains an
explicit expression for the kinetics of removal and
creation of bonds. The kinetic equation (9) can be
adapted to the specific properties of the bonds
occurring in a given material. Thus, if the network
consists of particles held together by colloidal
interaction forces one might prefer to use an expression derived from flocculation kinetics [43,44]
instead of the creation-term of equation (9).

A still more recently developed model [25,45] accounts for non-affine deformation of the network in the strained sample. The "slip" of liquid along network constituents is determined by the introduction of a factor $(1 - \xi)$ in the constitutive equation:

$$p'_{ik(n)} + \frac{\varepsilon}{G_n} p'_{ik(n)} \operatorname{tr} p'_n + \lambda_n \frac{\delta}{\delta t} p'_{ik(n)}$$

$$= \frac{2G_n\lambda_n}{1 - \xi} e^{(1)}_{ik} , \qquad (10)$$

together with equation (2) accounting for the spectral decomposition of the stress tensor. The result (10) is similar to equation (7) in that the rate of bond rupture is determined by the stored elastic energy, tr p. The parameters λ_n and G_n are obtained from the "equilibrium" relaxation spectrum, without shifting to shorter time as in equations (3) or (8). The parameters ξ and ε can in principle be derived from experiments. It is interesting to note that ξ determines the behaviour in simple shearing flow, whereas extensional flow measurements are required to determine ε.

The possibility to allow for non-affine deformations might be helpful in making a more realistic model. Extreme examples of such non-affine deformations are encountered in the testing for "juiciness" of meat or certain fruits. Another interesting feature of equation (10) is that it offers the possibility of direct experimental determination of the rates of destruction and of creation of network junctions [46], without *a priori* assumptions about the kinetic equation.

The material specificity enters the theory mainly through the kinetic equation. Testing of the various theories for their usefulness should be based on their predictive capacity, in the relation between rheological properties and structure (as this becomes manifest in the relaxation spectrum and the kinetics of rupture and creation of bonds). An illustration is given in Fig. 8, showing that an extension of the long-time end of the relaxation spectrum results in an increased toughness according to Marrucci's model. This is what one would expect, and this result offers the possibility of controlling toughness by proper modification of the long-term relaxation processes. The obvious way to do that is to decrease the mobility of the units constituting the network, either by making them bigger or by making the bonds between them stronger.

Much more work is required to determine the optimal form of the equations, and the proper rheological technique, for any given food product. Some possible areas of application of the theory will now be discussed.

7. INHOMOGENEITY

Most food products are inhomogeneous on a microscopic, and sometimes even on a macroscopic scale. Handling of such products by the consumer will normally result in inhomogeneous deformation. It can even be argued that the occurrence of stress-overshoot in a start-shear experiment is necessarily associated with non-uniform deformations: coherence in the sample is gradually lost in certain locations but not in other regions. The use of continuum theory for describing such deformations puts a limitation on the scale of the inhomogeneity: this should be small compared with the smallest dimension of the sample.

Inhomogeneity on a microscopic or colloidal scale is probably the rule rather than the exception in the kind of materials considered here: the network-forming units are not usually distributed uniformly in space, but a sample contains regions having a high concentration of such units, surrounded by regions in which the concentration is less than average.

In a material containing units subject to mutual attraction, the flocculated or gelled state will generally be inhomogeneous. This is now well established for flocculated particle suspensions [47-49]. For polymeric systems, the non-uniformity appears immediately from the opacity of certain gels. The phenomenon has been studied by optical techniques [50,

51] in transparent systems, and more recently by permeametry [52-54].

A relation between inhomogeneity and consistency is suggested by the well-known application of block co-polymers to produce "tough" plastics. In terms of the temporary network theory, this would mean that the state of aggregation affects the relaxation spectrum, in particular at the long-time end. Measurements by Weiss and Silberberg [54] on inhomogeneous gels seem to confirm this. Relatively small changes in the long-time spectrum may result in appreciable changes in behaviour, as exemplified by the effects of molecular weight distribution on shear thinning, drag reduction, and rupture in elongational flow. In many food products, the degree of inhomogeneity can easily be controlled by a judicious choice of process conditions, and this might offer a route for improved control of product properties.

8. ADDITIVES

A popular method for modifying consistency is by adding "thickeners", stabilizers or gelling agents. These are often polysaccharides, which are available in a large variety as natural products. Their behaviour depends strongly on the molecular architecture [55] which allows large variations in chain rigidity to occur between the various species. Several different crosslinking mechanisms may be operative in forming gels in water; typical examples are double helices in carrageenans, multiple ionic bonds in alginates and in certain pectins, and hydrophobic bonds in cellulose derivatives. The gelling behaviour can further be modified by varying the pH or the salt content of the solution, and it is sometimes affected by the presence of other components in the system. In particular, mixtures of different polysaccharides may exhibit a behaviour that is different from that of each of the separate components.

It is obvious that the contribution of a polysaccharide to the rheology of a food product cannot be described by the well-known theories of polymer solutions.

Very few examples of complete rheological investigations of such products have been published; a notable
exception is the work on dough [56,57]. No attempt
was made to relate the results of this work to a model
(i.e. constitutive equation) because no suitable model
was available at that time.

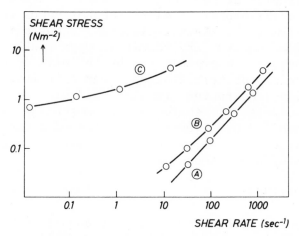

*Fig.9 Flow properties of stabilized suspension of
carbon particles in water (line A); volume
fraction 0·10. Curve B: 1mg per m² surface
area Polyvinylalcohol added. Curve C same
composition as B, but processed to form
polymer bridges between particles.*

   In food products containing dispersed particles or
droplets, additives in the form of biopolymers may
affect the consistency by modifying the properties of
the particle network. Colloidal interaction forces
between dispersed particles may lead to the formation
of a temporary network [9,20]. Fig. 9 illustrates how
the properties of the network are altered drastically
by allowing interaction to take place between layers
of adsorbed polymer at the particle surfaces. Curves
B and C relate to dispersions having the same overall

composition. System B has a polymer layer adsorbed
on every particle; the layer prevents particles from
approaching each other to sufficiently small distances
for network formation to be possible. System C was
prepared by mixing uncovered particles with a suspension of polymer-covered particles, such that polymer
molecules could form bridges between neighbouring
particles. The difference in consistency is spectacu-
lar, and probably far bigger than can be obtained by
any other method in a system of constant composition.
The phenomenon relies on the fact that polymer adsorption is effectively irreversible, and hence the system
remembers indefinitely how it was prepared.

9. CONCLUSION

The use of Rheometry in the food industries requires
the availability of a model theory, for relating
rheological parameters to processes occurring on a
molecular or colloidal scale in the deformed material.
Such a model, taking the form of a constitutive
equation, has recently been developed on the basis of
the temporary network theories. Examples are given
of possible areas where application of the model might
result in improved control of product consistency.

REFERENCES : CHAPTER 4

[1] G.G. Birch, J.G. Brennan and K.J. Parker,
     Sensory Properties of Foods, Applied Science
     Publishers, 1977

[2] P. Sherman, Industrial Rheology, Academic Press,
     1970

[3] L.L. Thurstone, Multiple Factor Analysis,
     Chicago, 1947

[4] W.M. Breene, J. Texture Stud. 9 (1978) 77

[5] P. Howgate, ref. [1] p. 249

[6] R.A. Segars, R.G. Hamel and J.G. Kapsalis,
     J. Texture Stud. 8 (1977) 433

[7] J.G. Oldroyd, Proc. R. Soc. London A218 (1953)
     122

[8] J.D. Ferry, Visco-elastic Properties of Polymers,
     John Wiley & Sons, Inc., New York, 1970

[9] M. van den Tempel, J. Colloid Sci. 16 (1961) 284

[10] J.M.P. Papenhuijzen, Rheol. Acta 11 (1972) 73

[11] J. Lyklema, M. van den Tempel and T. van Vliet,
     Rheol Acta 17 (1978) 525

[12] J.D. Huppler, E. Ashare and L.A. Holmes, Trans.
     Soc. Rheol. 11 (1967) 159, 181

[13] F.H. Gaskins, J.G. Brodnyan and W. Philippoff,
     Trans. Soc. Rheol. 13 (1969) 17

[14] P.J. Carreau, Trans. Soc. Rheol. 16 (1972) 99

[15] H.C. Booy, Ph.D. thesis, Univ. of Leyden,
     "Effect of superimposed shear flow on dynamic
     properties of polymeric liquids", 1970

[16] N.N. Moshenin and J.P. Mittal, J. Texture Stud. 8 (1977) 395

[17] J.M.P. Papenhuijzen, Rheol. Acta 10 (1971) 493, 503

[18] R.I. Tanner and G. Williams, Rheol. Acta 10 (1971) 528

[19] J. Mewis and G. Schoukens, Faraday Discuss. nr 65, 1978

[20] G. Schoukens and J. Mewis, J. Rheol. 22 (1978) 381

[21] J.S. Dodge and J.M. Krieger, Trans. Soc. Rheol. 15 (1971) 589

[22] D.W. de Bruijne, N.J. Pritchard and J.M.P. Papenhuijzen, Rheol. Acta 13 (1974) 418

[23] D.M. Binding, J.M. Davies and K. Walters, J. Non-Newtonian Fluid Mechanics 1 (1976) 259, 277

[24] P.D. Dokic and L.J. Djakovic, J. Colloid Interface Sci. 51 (1975) 373

[25] N.P. Thien and R.I. Tanner, J. Non-Newtonian Fluid Mechanics 2 (1977) 353

[26] T.L. Smith, Pure Appl. Chem. 23 (1970) 235

[27] H.M. James and E. Guth, J. Chem. Phys. 11 (1943) 455

[28] R.D. Andrews, A.V. Tobolsky and E.E. Hanson, J. Appl. Phys. 17 (1946) 352

[29] A.V. Tobolsky et al. J. Chem Phys. 14 (1946) 80, 93, 113

[30] P.J. Flory, Trans. Faraday Soc. 56 (1960) 722

206

[31] A.S. Lodge, Trans. Faraday Soc. 52 (1956) 120

[32] B.J. Meister, Trans. Soc. Rheol. 15 (1971) 63

[33] W.W. Graessley, J. Chem. Phys. 43 (1965) 2696

[34] R.I. Tanner and J.M. Simmons, Chem. Eng. Sci. 22 (1967) 1803

[35] G. Marrucci and G. Astarita, Rheol. Acta 13 (1974) 754

[36] G. Marrucci, G. Titomanlio and G.C. Sarti, Rheol. Acta 12 (1973) 269

[37] D. Acierno, F.P. La Mantia, G. Marrucci and G. Titomanlio, J. Non-Newtonian Fluid Mechanics, 1 (1976) 125

[38] D. Acierno, F.P. La Mantia, G. Marrucci, G. Rizzo and G. Titomanlio, J. Non-Newtonian Fluid Mechanics 1 (1976) 147

[39] D. Acierno, F.P. La Mantia and G. Titomanlio, Rheol. Acta 15 (1976) 642

[40] D. Acierno, F.P. La Mantia and G. Marrucci, J. Non-Newtonian Fluid Mechanics 2 (1977) 271

[41] D. Acierno, F.P. La Mantia, B. de Cindio and L. Nicodemo, Trans. Soc. Rheol. 21 (1977) 261

[42] W.W. Graessley, W.S. Park and R.L. Crawley, Rheol. Acta 16 (1977) 291

[43] M. van den Tempel, Emulsion Rheology (Ed. P. Sherman); p. 1, Pergamon Press, 1962

[44] E.R. Ruckenstein and J. Mewis, J. Colloid Interface Sci. 44 (1973) 532

[45] N.P. Thien, J. Rheol. 22 (1978) 259

[46] P.F. van der Wallen Mijnlieff and
R.J.J. Johgschaap, Private communication

[47] A.S. Michaels and J.C. Bolger, Ind. Eng. Chem.
1 (1962) 153

[48] T.G.M. van der Ven and R.J. Hunter, Rheol. Acta
16 (1977) 534

[49] M. van den Tempel, J. Colloid Interface Sci.
(to be published) 1979

[50] K.L. Wun, G.T. Feke and W. Prins, Faraday
Discuss. 57 (1974) 146

[51] T.T. Tanaka, Phys. Rev. A17 (1978) 763

[52] P.F. Mijnlieff and W.J.M. Jaspers, Trans.
Faraday Soc. 67 (1971) 1837

[53] P.F. Mijnlieff and F.W. Wiegel, J. Polym. Sci.,
Polym. Phys. Ed. 16 (1978) 245

[54] N. Weiss and A. Silberberg, Brit. Polym. J.
(1977) 144

[55] D.A. Rees, Biochem. J. 126 (1972) 257

[56] N.W. Tschoegl, J.A. Rinde and T.L. Smith,
Rheol. Acta 9 (1970) 223

[57] J.R. Smith, T.L. Smith and N.W. Tschoegl,
Rheol. Acta 9 (1970) 239

CHAPTER 5

MOLTEN POLYMERS

By

J.L. White

CONTENTS

1. INTRODUCTION

In general, rheological measurements on polymer melts
and elastomers face great difficulties. Experiments
must be carried out at elevated temperatures and sometimes at elevated pressures. In many cases special
precautions with the environment or with the removal
of impurities must be made. To proceed, one must
understand not only rheological principles but the
peculiar characteristics of the systems to be investigated.

The instruments used to characterise these materials
are as is the case with most classes of industrial
materials of two general types. There are scientific
instruments intended to make basic measurements of
rheological properties and rougher simpler apparatus
which is primarily used for industrial quality control.
Both of these are of concern to us.

Our organisation of this chapter will be to first
review the major aspects of polymer structure and
processing condition chemical stability. We will then
turn to a classification of flows which may be studied
in rheometers and discuss each major type of instrument
in turn. We consider first basic instruments. The
correlation of rheological properties with molecular
structure is then considered. Finally we turn to

industrial control apparatus.

2. POLYMER STRUCTURE AND INDUSTRIAL POLYMER SYSTEMS

2.1 General

We must first begin with some consideration of the types of materials we are concerned with. We must of necessity be concise. Polymers are in general high molecular weight long chain molecules. In the simplest case, a polymer consists of a single repeating unit [1]

$$A A A A A A A A A . . . \tag{I}$$

Many of the most important polymers are of this type.

The most important type of linear polymers are vinyl polymers with the repeating unit [1,2]:

$$+ CH_2 - \overset{*}{C}H - \underset{\underset{R}{|}}{} + \tag{II}$$

which includes polyethylene (R $\equiv$ H), polypropylene (R $\equiv$ -CH), polystyrene (R $\equiv$ phenyl, - 0), polyvinyl chloride (R $\equiv$ Cl). They have been shown to possess mainly head to tail structures. The carbon atom with the asterisk is asymmetric. Regularities in its geometric configuration make the polymer crystallise. If all the carbons possess the same configuration, the polymer is said to be isotactic [2]. Commercial polypropylene is primarily isotactic while commercial polystyrene is without order and is said to be atactic. It cannot crystallise and vitrifies to form a glass. Vinyl polymers are used commercially primarily as plastics though some polypropylene is melt spun to form fibre.

Another important class of polymers are polydienes which are formed from the monomer [1]:

$$CH_2 = \underset{\underset{X}{|}}{C} - CH = CH \tag{III}$$

and generally have the mixed structural units

$$\left[CH_2 - \underset{\underset{X}{|}}{C} = CH - CH_2\right] / \left[CH_2 - \underset{\underset{\underset{CH_2}{\|}}{C-X}}{CH}\ \dot{-}\ \right] / \left[CH_2 - \underset{\underset{\underset{CH_2}{\|}}{CH}}{\overset{\overset{X}{|}}{C}} - \right]$$

    1,4 addition   3,4 addition   1,2 addition

$$\ldots \quad (IV)$$

The double bond may be <u>cis</u> or <u>trans</u>. Typical important polymers of this type are polybutadiene (X ≡ H),
polyisoprene (X ≡ CH_3) and polychloroprene (X ≡ Cl).
The commercial dienes are used as elastomers and
generally have predominantly the 1,4 addition form.

Mention should also be made of condensation polymer
[1,2] including notably polyamides

$$\left[NH - R - NH - \overset{\overset{O}{\|}}{C} - R' - \overset{\overset{O}{\|}}{C} \right] \qquad (Va)$$

or

$$\left[NH - R'' - \overset{\overset{O}{\|}}{C} \right] \qquad (Vb)$$

and polyesters

$$\left[O - X - O - \overset{\overset{O}{\|}}{C} - X' - \overset{\overset{O}{\|}}{C} \right] \qquad (VIa)$$

$$\left[O - X'' - \overset{\overset{O}{\|}}{C} \right] \qquad (VIb)$$

The most important of such polymers are nylon-6 (R'' ≡
$(CH_2)_5$), nylon-66 (R ≡ $(CH_2)_6$, R' ≡ $(CH_2)_4$), and polyethylene terephthalate (X ≡ (CH) , X' ≡ - O -).

Polyamides and polyesters are used as fibres, films and plastics.

Polymer chains may be branched as well as linear, i.e. they may possess structures of form [1]:

```
A A A A A A A A A A A A A A A A A A A
A       A           A   A   A           A
A       A           A   A   A           A
            A A A           A
                            A
                            A
                            A                   (VII)
```

The branches may be either 'short' or 'long' and the branches themselves may have branches.

In addition polymer chains may be crosslinked into three-dimensional structures [1]:

```
A A A A A A A A A A A A
        A           A
        A           A
        A           A
        A           A
A A A A A A A A A A A A A A A A
            A           A
            A           A
            A           A
            A           A
    A A A A A A A A A A A A A A      (VIII)
```

This causes the polymer system to be unable to flow under the action of stresses and to respond as a solid rather than a fluid.

Another class of polymers are copolymers which have more than one structural unit. The arrangement of the units in a copolymer chain may vary considerably. The units may be arranged in random

```
    A A B A B A B B B A B A B A A A A B B A          (IX)
```

alternating

A B A B A B A B A B A B A B A B A   (X)

or block structures such as:

AB

. . . . . . A A A A A A A ⦂ B B B B B B . . . . .

(X1)

or

ABA

A" A A A - - - A A ⦂B B B - - - - B B B ⦂ A A . . . . A A A A'

(XII)

All types of intermediate structures between (IX), (X), (XI) and (XII) are possible. The random, AB and ABA type block copolymers are produced commercially.

The copolymers indicated above are linear. It is of course possible to have branched copolymers with random, alternating, block or intermediate structures. Of special interest are the graft copolymers with structure [4]

```
A A A A A A A A A A A A
    B           B
    B           B
    B           B
    B           B              (XIII)
```

where B chains grow off an 'A' backbone.

2.2 Molecular Weight and Molecular Weight Distribution
      [3 - 7]

Commercial polymers generally contain a distribution of molecular weights. This distribution is frequently specified in terms of average molecular weights which are moments of the distribution. We may define the number, weight, z and z+1 average molecular weights

M_n, M_w, M_z and M_{z+1} by:

$$M_n = \frac{\sum N_i M_i}{\sum N_i} , \tag{1a}$$

$$M_w = \frac{\sum N_i M_i^2}{\sum N_i M_i} , \tag{1b}$$

$$M_z = \frac{\sum N_i M_i^3}{\sum N_i M_i^2} , \tag{1c}$$

$$M_{z+1} = \frac{\sum N_i M_i^4}{\sum N_i M_i^3} . \tag{1d}$$

The ratio M_w/M_n is frequently used as a polydispersity index.

Methods of measurement of molecular weight and its distributions are discussed in numerous monographs. This is reviewed concisely by Flory [3] (early work to 1952) and Schultz [4] and in more detail in monographs by Tanford [5], Cantow [6] and Morawetz [7]. The major methods of absolute determination of molecular weights are colligative properties (osmotic pressure, freezing point depression) and light scattering. The most commonly used 'relative' method of characterising molecular weight is dilute solution (intrinsic) viscosity. Molecular weight distributions are most usually estimated with chromatographs operating on a molecular sieve principle (gel permeation chromatography, GPC).

The absolute levels of molecular weight of commercial polymers vary considerably but are in any case in excess of 10,000. Below molecular weights of 10 –

50,000, depending on polymer type, the material has the mechanical properties of a low molecular weight wax. Only above this range of values do polymers achieve the unique combination of properties which make them of commercial importance. Condensation polymers such as polyesters and polyamides tend to be in the lower range of molecular weights with values of the weight average molecular weight M_w of 15 - 40,000 while vinyl polymers and polydienes generally have values of M_w in excess of 200,000.

The breadths of molecular weight distributions vary considerably among polymers. Anionic polymerisation is capable of producing very narrow distribution polymers. The standards for gel permeation chromato-graphs developed by this technique have values of M_w/M_n less than 1.1. Commercial polybutadienes and butadiene-styrene copolymers produced by this mechan-ism often have M_w/M_n of 1.5 to 2 [8]. The kinetics of condensation polymerisation generally results in polyesters and polyamides having values of M_w/M_n close to 2 [3,4]. Commercial polystyrenes generally have values of M_w/M_n in the range of 2.5 to 4 [8] and polypropylenes of 5 to 10 [9]. Polyethylene poly-dispersity indices M_w/M_n can be as low as 5 - 6 but are frequently in the range 20 - 30. Emulsion poly-merised butadiene-styrene copolymers appear to have polydispersity indices of order 10 [8].

2.3 Primary Transition Temperatures [10 - 13]

The primary transition temperature determines when a polymer transforms from a solid to a fluid state capable of processing. In some cases this is a first order transition, a crystalline melting point, T_m, in others it is a 'glass transition' temperature T_g. Values of T_m and T_g for important commercial polymers are listed in Table 1.

TABLE 1

TRANSITION TEMPERATURES AND OPERATING TEMPERATURES FOR
POLYMERS

Polymer	T_m (Melting Temperature)	T_g (Glass transition temperature)	Normal Melt Processing Temperature Range
High Density Polyethylene (HDPE) Low Density Polyethylene (LDPE)	140°	∿–100°	160–240°C
Isotactic Polypropylene (PP)	165°	∿– 15°	180–240°C
Polystyrene (PS)		100°	180–240°C
cis-polyisoprene (NR)	30°	– 70°	90–110°C
cis-polybutadiene (BR)	0°	–110°	90–110°C
Butadiene-styrene copolymer (23.5% styrene) (SBR)		– 55°	90–110°C
Polyethylene terephthalate (PET)	265°	70°	275–290°C
Nylon-66	265°	40°	275–290°C
Nylon-6	220°	40°	230–260°C

Not all polymers crystallise. The process of crystallisation is limited as indicated earlier to regular
structures such as tactic hydrocarbons and linear aliphatic polycondensates. Random copolymers do not
crystallise. Among important homopolymers which crystallise are polyethylene, polypropylene and nylon-66.

Polymers which vitrify to form glasses include polystyrene and polymethyl methacrylate. Polycarbonate and polyethylene terephthalate crystallise very slowly and usually form glasses. However, under extreme conditions they crystallise. Nylon-6 is peculiar in that it usually supercools to a glass, absorbs moisture and then crystallises [14].

Elastomers exhibit transition temperatures considerably below room temperature usually in the range -50°C to -110°C, These materials are neither fabricated nor applied in this temperature range. Some elastomers with regular structures crystallise and exhibit T_m in a range where they are used. Cis-1,4 polyisoprene (natural rubber) has a value of T_m about 30°C and cis-1,4 polybutadiene around 0°C.

The value of T_m is frequently suppressed in practice by finite cooling rates. The crystallisation rates at T_m are very small and achieve a maximum at temperatures approximately halfway between T_m and T_g. At T_g and below the rates are zero. Thus polypropylene ($T_m = 165^{\circ}$, $T_g = -15^{\circ}$C) frequently crystallises in the range $90-110^{\circ}$C [12]. Polyethylene terephthalate, as mentioned earlier, usually crystallises so slowly that it cools through the $T_m - T_g$ range and forms a glass.

Crystallisation rates are greatly enhanced by the action of applied stresses. Thus cis-1,4 polyisoprene, usually amorphous at 20°C, crystallises on stretching [11]. Crystallisation rates in polydienes and polyolefins are well known to be increased by orders of magnitude by applied stresses [11,12,15]. Polyethylene terephthalate crystallises in melt spinning operations when subjected to high stresses and rates of drawdown [16].

In typical polymer processing operations, polymers are subjected to both supercooling and stress induced crystallisation. The results on each material depend upon the individual material characteristics.

2.4 Changes in Polymer Structure during Flow

Chemical changes can and do occur in the flow of
polymer melts. The character of these changes varies.
It would appear that some are due to purely thermal
effects. Others are due to chemical reactions with
impurities brought on by elevated temperatures.
Another class of chemical change is associated with
degradation or breakage of polymer chains through
applied stresses. Often the detailed reactions can
be complex and involve all three of the above aspects
with the relative proportion varying with temperature.
It is worthwhile to consider the situation for
different particular polymers.

Condensation Polymers. Polymers of this type, especially polyesters, are highly subject to hydrolysis by
the presence of small amounts of moisture, i.e.

$$- O - X - O - \overset{\overset{\textstyle O}{\|}}{C} - X' - \overset{\overset{\textstyle O}{\|}}{C} - + H_2O \rightarrow$$

$$\| O - X - OH + HO - \overset{\overset{\textstyle O}{\|}}{C} - X' - \overset{\overset{\textstyle O}{\|}}{C} - \qquad (XIV)$$

The forward rate of this reaction increases rapidly
with temperature and is very significant at the processing temperatures used for example in polyethylene
terephthalate $(275 - 290^\circ C)$. Extreme care must always
be taken in drying condensation polymers, especially
polyesters. By reaction (XIV) there is a tendency for
polymers to form an equilibrium molecular weight
determined by the moisture content, i.e. the molar
concentration of H_2O through

$$\frac{[H_2O]\,[-\, O\overset{\overset{\textstyle O}{\|}}{C} -]}{[COOH]\,[OH]} = K . \qquad (2)$$

Polyamides exhibit tendencies to gel at high temperatures as well as hydrolyse and one must be careful to avoid either problem. This is discussed by Pezzin and Gechele [12] for nylon-6. The gelling effects in nylon-66 are more serious perhaps due to the necessity of using higher operating temperatures.

Vinyl Polymers. Commercial polyolefins, notably polypropylene, also exhibit chemical instability during melt flow. The presence of oxygen is necessary or greatly accelerates the rate of breakdown of polypropylene. The particular instability of polypropylene is associated with the removal of the tertiary hydrogen, e.g.

$$- CH_2 - \underset{\underset{CH_3}{|}}{CH} - CH_2 - \underset{\underset{CH_2}{|}}{CH} - \rightarrow - CH_2 - \underset{\underset{CH_3}{|}}{\overset{\bullet}{C}} - CH_2 - \underset{\underset{CH_3}{|}}{CH} -$$

$$\rightarrow - CH_2 - \underset{\underset{CH_3}{|}}{C} = CH + \cdot \underset{\underset{CH_3}{|}}{CH} - \qquad (XV)$$

This behaviour has been considered in some detail by Kowalski and his coworkers [18,19] in the patent literature. The random degradation of polypropylene not only lowers molecular weight but tends to produce products with M_w/M_n of 2 which substantially narrows the distribution which originally possesses an M_w/M_n of 10. Temperatures of about 270°C and higher are required. It is believed that processes of this type are used commercially to produce lower molecular weight grades of polypropylenes.

Polyvinyl chloride can degrade and release HCl at elevated temperatures

$$- CH_2 - \underset{\underset{Cl}{|}}{CH} \rightarrow - CH = CH - + HCl \qquad (XVI)$$

if not properly stabilised. The HCl in small quantities will corrode machinery and in large quantities can injure workers' health.

Mechanochemical Degradation. Another type of flow induced chemical change in polymers is direct breakup of applied stresses. This effect was first discovered and commercially exploited by Thomas and Hancock [20, 21] from about 1820 with natural rubber. It was not, however, until the present century that the efforts of Busse [22,23] and Watson and his coworkers [24-26] explained the mechanism of the behaviour and the key role played by oxygen. This is briefly summarised below:

$$R - R' \xrightarrow{\text{stress}} R. + R'.$$

$$R. + O_2 \longrightarrow RO_2^{\bullet} \quad \text{(relatively stable)}$$

$$R. + A \longrightarrow RA. \quad \text{or} \quad RO_2^{\bullet} + A \longrightarrow RO_2A \quad \text{(XVII)}$$

Oxygen acts to stabilise the free radicals and prevent them from recombining. The reaction may be hastened by the addition of efficient free radical acceptors, A [26].

While the process is most effective with cis-1,4 polyisoprene, it is known to occur with a wide range of polymers [27-31]. In masticated blends, one may form graft copolymers from interaction between radical segments and polymer chains of other species [32]. Polymers dissolved in monomers can upon mastication break down to form free radicals which may polymerise the monomer to form a block copolymer [33-36].

2.5 Multiphase Systems and Compounds

Many polymer systems used industrially consist of more than a single polymeric component and phase. We mean here more than the obvious minor amounts of stabilisers, antioxidants, etc. generally added. Many industrial polymer systems consist of physical blends of different polymers or physical mixtures of two different components with a thick interface of a graft copolymer. This is the case for example of rubber modified (high impact) polystyrene and ABS resins [37, 38]. In still other cases, polymer systems may contain

large quantities of solid particulates which are added
as a reinforcing agent to improve product properties
(e.g. carbon black silica or glass, aramid or cellulose fibres) or as a filler (clay) [39,40]. Many
polymeric products have complex recipes involving
several polymer components, different solid particulate additives plus organic oils. In rubber compounds
one usually has crosslinking agents (sulphur plus
acceleration) present as well [39].

Mechanical blends of different polymers, not
unexpectedly, generally possess similar rheological
properties [41,42] and flow in a manner similar to
single polymers alone. (But see reference [43]).
However, the addition of significant quantities of
solid particulates can cause major changes in flow
characteristics. Carbon black [44-46], calcium
carbonate [46], talc [47], titanium dioxide [46,48]
and other small particulates result in producing
'particle-polymer' networks which behave as gels. In
the case of compounds with brittle particles such as
glass fibres, rheological properties may change during
flow due to fibre damage [49,50]. More complex
systems such as rubber modified plastics with crosslinked rubber particles and graft copolymer (e.g. ABS
resins) interfaces also tend to show yield values
[51-53].

3. GENERAL OBSERVATIONS OF KINEMATICS AND PROCESSING

3.1 Types of Flows and their Stress Fields [59]

Various types of rheological instruments may be used
to investigate the response of polymer melts during
flow. These may be classified according to the kinematics of the flow (cf. Chapter 1). Generally we may
divide the flows of interest into shear and elongational flows. We shall describe each of these in turn.
We also consider the stress responses to these flows
which can be represented in terms of the tensor which
can be resolved into pressure p plus 'extra stress'
p'_{ij} [54]

$$p_{ij} = -p\delta_{ij} + p'_{ij} .$$
(3)

We begin by establishing a '123' Cartesian coordinate system. A shear flow may be specified by

$$v_1 = qx_2 , \qquad v_2 = v_3 = 0 , \tag{4a}$$

where q is a velocity gradient or shear rate. The primary stress response to this flow is the shear stress τ which is related to the shear rate q by the viscosity η

$$\tau = \eta q , \tag{4b}$$

though non-isotropic normal stresses p_{11}, p_{22} and p_{33} usually develop.

In elongational flows we have

$$v_1 = k_1 x_1 , \qquad v_2 = k_2 x_2 , \qquad v_3 = k_3 x_3 , \tag{5a}$$

where the k_j are elongation rates which are restricted by

$$k_1 + k_2 + k_3 = 0 . \tag{5b}$$

Different special classes of elongational flows may be distinguished. First there is <u>uniaxial extension</u> for which

$$k_1 = k , \tag{6a}$$

$$k_2 = k_3 = - \tfrac{1}{2} k .$$

Here stresses p_{ii} develop but p_{11} is the primary stress of interest. p_{22} and p_{33} are generally equal to the pressure of the environment. p_{11} is related to k through an elongational viscosity η_E:

$$p_{11} = \eta_E k . \tag{6b}$$

We may also consider multiaxial stretching and flows. We may similarly define <u>biaxial extension</u> where

$$k_1 = k_2 , \qquad k_3 = - 2k_1 . \tag{7a}$$

The stresses developed are p_{ii} with p_{11} equal to p_{22}. In planar extension there is no deformation in one direction.

$$k_1 = -k_2 , \quad k_3 = 0 . \tag{7b}$$

3.2 Viscous Dissipation Heating

Polymer melts possess very high viscosities compared to other industrial fluids. The viscosities may range from 100 to 100,000 Pa.s. (1000 to 1,000,000 poise). One result of this is that relatively large amounts of power must be applied to operate rheological instruments. This is converted to heat during the operation of the rheometer. The rate of generation of heat Φ is given by [55]:

$$\Phi = \tau q . \tag{8a}$$

The maximum (adiabatic) temperature rise may be determined from the solution of [59]

$$\rho c \left. \frac{dT}{dt} \right|_{res} = \tau q = \eta q^2 . \tag{8b}$$

A fluid with a viscosity of 5000 Pa.s which is sheared at a shear rate q of 100 $\sec^{-1}$ shows an adiabatic temperature rise of 12°C per second. Obviously this high level of temperature rise is somewhat moderated by heat conduction, but its importance should be clear.

In general, analysis of temperature profiles developed due to viscous heating requires the solution of the full energy equation which may be expressed (in Cartesian tensor form) as [59]

$$\rho c \left[\frac{\partial T}{\partial t} + v_m \frac{\partial T}{\partial x_m} \right] = \bar{k} \frac{\partial^2 T}{\partial x_m \partial x_m} + P_{ij} \frac{\partial v_i}{\partial x_j} , \tag{9}$$

where $\bar{k}$ is thermal conductivity. The first term on the left hand side accounts for transient rise in energy (heat) content and the second term for heat convection. The first term on the right hand side

accounts for heat conduction and the second is the
viscous heat generation. Solutions of this equation
for different viscometric geometries have been
reported by various investigators [56-59].

Certain general comments are useful. In the steady
state the terms on the left hand side of Equation (9)
may be neglected. The magnitude of the temperature
rise ΔT_{rise} will be determined by the dimensionless
group representing the ratio of the viscous dissipation to the heat conduction term. This is known
as the Brinkman number [55]. Thus

$$\Delta T_{rise} \sim N_{Br} = \frac{(\text{viscous dissipation})}{(\text{heat conduction})} = \frac{\eta \left(\frac{U}{L}\right)^2}{\bar{k} \left(\frac{\theta}{L}\right)^2}$$

$$= \frac{\eta U^2}{\bar{k}\theta} = \frac{\eta q^2}{\bar{k}\theta} L^2 \,, \tag{10}$$

where U is a characteristic velocity and L a characteristic length, generally the instrument gap size.
Clearly at fixed q, Brinkman number and ΔT_{rise}
increase with the square of L. One minimises ΔT_{rise}
by keeping L small.

3.3 Polymer Fabrication Technology [60-67] (Figure 1)

A major purpose of industrial rheological measurements is the interpretation of flow in polymer processing operations and the forming of products from
polymer systems. This includes the technology of the
synthetic fibre, plastics and rubber industries.

In most processing operations, the polymer is pumped
through the action of a rotating screw, i.e. a screw
pump (Figure 1a). Generally pellets or cold slabs are
added to the hopper of the extruder and these are
transported by the motion of the screw and softened
or melted by the combined action of heat transfer
from a heated barrel and viscous heat generation
within the screw channel. The transport in the fully

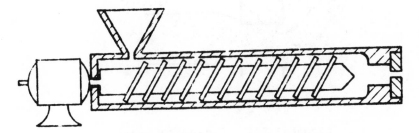

FIGURE 1 (a). Screw Extruder.

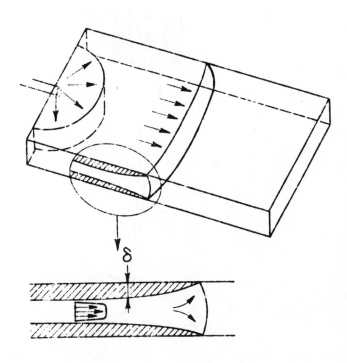

FIGURE 1 (b). Injection Mold Filling.

228

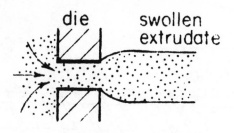

FIGURE 1 (c). Extrudate Swell

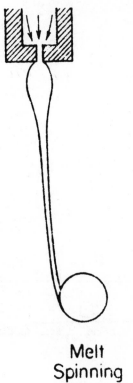

Melt
Spinning

FIGURE 1 (d). Melt Spinning.

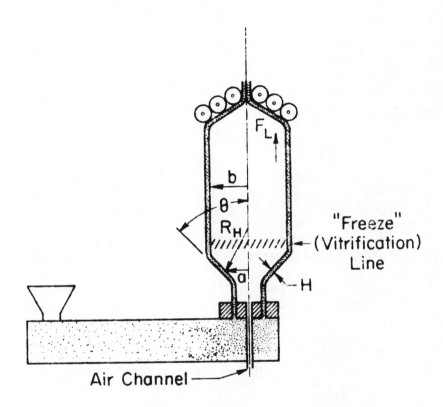

Tubular Film Extrusion

FIGURE 1 (e). Tubular film extrusion.

melted polymer in the metering region of the screw
extruder, which is most distant from the hopper, is
basically a shearing flow. Analysis of transport in a
screw involves understanding solid transport, melting
and rheological behaviour. However, only shear flow
behaviour need be understood and applied.

The output of the extruder screw generally proceeds
through a die and thence into air where it may be subsequently treated. Flow of the melt in the die as
well as in the metering region of the screw is dominated by the contact and adhesion of the melt with the
solid walls. The motion of the melt is determined by
the application of pressure or the relative motion of
the walls. This causes the motions to be basically
shearing in character. An exception is in the region
where the melt converges into the narrow die channel.
Here there may be mixed shearing and uniaxial stretching motions. Near the centreline of the die, the
shearing behaviour disappears because of symmetry and
uniaxial stretching will dominate. In extrusion of
pipes or profiles, the flow within the die will be
predominantly shearing and the shear flow rheological
behaviour will be primarily needed for interpretation.

Following the die exit, three different types of
subsequent histories may occur. In *injection moulding*,
the melt is injected into a mould. Again the walls
dominate and shearing flow occurs. As in the extruder
screw, the flow may be highly non-isothermal. This is
due to cooling of the melt by cold mould walls and
heating by viscous dissipation (Figure 1b).

In *profile and pipe extrusion* processes the melt
emerges into the air where it undergoes a 'recovery'
process due to the release of the applied stresses
along the die wall. The resulting swell can result
in changes of the profile shape (Figure 1c).

In many processing operations tensions are applied
to the emerging melt in order to change its shape. We
may classify the drawdown in *melt spinning* of fibres
as such a process as well as the combined drawdown and
blow up of the annular melt sheet during *tubular film*

extrusion. In some processes such as *blow moulding* and *thermoforming* the emerging melt is allowed to undergo an unstressed 'recovery' period as in profile extrusion, but within a short time interval it is exposed to applied tensions which stretch it into a new shape. Melt spinning involves a uniaxial extension process, while tubular film extrusion, flow moulding and thermoforming involve multiaxial stretching. Understanding of the rheological response in these flows requires the study of elongational flow behaviour. It is also to be noted that all of these elongational flow processing operations are highly non-isothermal because of cooling of the emerging melt by the environment (Figures 1d, 1e).

Processing problems vary with the type of process. For design of 'internal flows' as in extrusion and injection moulding, the shear viscosity and its shear rate and temperature dependence are required. In extrusion one also has problems with varying extrudate swell levels and extrudate distortion which seem related to melt elasticity and 'memory'. Fibre and film extrusion have beyond this a swollen or distorted extrudate problem, fluctuations in diameter or thickness related to system response to input disturbances (and sometimes the 'draw resonance instability' [64, 65]). This may be traced to the elongational flow properties of polymer melts and the cooling conditions during drawdown.

4. SHEARING FLOW INSTRUMENTS

4.1 General

In this section we shall consider basic scientific instruments, which have been used to measure rheological properties in shear flow. Measurements in shear flow were the first rheological data to be reported on polymer melts and indeed until relatively recently the only such studies reported.

4.2 Sandwich Viscometer (Figure 2)

The simplest of viscometers used for polymer melts

are the parallel plate and sandwich instruments described by Zakharenko et al [68], Middleman [69] and
later authors [70,71]. In this instrument a very
viscous fluid, generally an elastomer, is sheared
between two parallel steel plates or in a sandwich
between three plates. The shear rate is

$$q = \frac{V}{H} \, , \tag{11}$$

where V is the velocity of the moving member and H
the interplate distance.

FIGURE 2. Sandwich Viscometer

If the moving member is programmed at a constant
velocity, a constant shear rate is obtained. The
applied force, F, may be determined as a function of
time. This yields after normalisation with area A,
i.e.

$$\tau = \frac{F}{2A} \, , \tag{12}$$

the shear stress as a function of time.

Transient start up shear stresses are determined
which if the material proves stable between the plates
give at long times the steady state stress. If this
is achieved, the viscosity may be determined as a
function of shear rate. In many cases, rather large
times and apparatus shear strains are required. Slippage often occurs in elastomer systems under such
conditions.

The advantage of this type of instrument is the
ability to achieve very low shear rates for materials
which generally exhibit slippage in non-pressurised
rotational instruments such as those based on the
cone-and-plate geometry.

Generally, in order to avoid slippage of elastomeric
test specimens, the contact surfaces are best knurled
and the system 'pressurised' at the beginning of
experiments [70].

4.3 Coaxial Cylinder   (Figure 3)

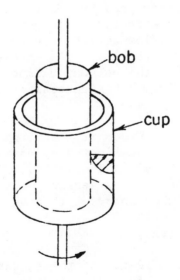

FIGURE 3. Coaxial Cylinder Viscometer.

The coaxial cylinder viscometer is one of the most important of rheological-characterisation instruments. It is of course one of the oldest of viscometers, having been developed and widely used in the late 19th century. The coaxial cylinder may involve a rotating inner cylinder (bob) or outer cylinder (cup).

This instrument was first applied to polymer systems by Mooney [72] in 1936 who reported the first true viscosity/shear rate data obtained on such a material (natural rubber) to appear in the literature. Few investigators have used this instrument since the time of Mooney. Philippoff and Gaskins [73] reported experimental studies in the mid 1950's using a coaxial cylinder viscometer with polyethylene. More recently, Cogswell [74] has used this type of instrument to study viscosity under conditions of controlled pressure.

There are various experimental problems reported with use of this instrument. One of the most striking which is reported by Mooney is due to the Weissenberg normal stress effect [75]. It must be remembered that the coaxial cylinder geometry is the classical one which exhibits this effect. Large normal stresses are developed in shear flow between the bob and cup and these give rise to "rod climbing". Mooney found that with natural rubber the polymer climbed out of the apparatus during shearing. This necessitated pressurisation at both ends.

The theory of viscosity measurement in this instrument has been investigated by various researchers dating to the 1920's [54,76-80]. The shear stress at the surface of the sensing member is simply related to the couple or torque C through

$$C = 2\pi r^2 L \tau(r) \ , \tag{13}$$

where L is the length of the sensing member. The shear rate q is more difficult to determine. If the cup is narrow,

$$q = \frac{r_i \Omega}{r_o - r_i} = \frac{\Omega}{\left(\dfrac{r_o}{r_i}\right) - 1} \quad , \tag{14}$$

where Ω is the angular velocity of the rotor and r_i and r_o are the radii of the inner and outer cylinders. Krieger, who has expended more effort [78-80] on this problem than any other investigator, notes in his most recent study [80] for the case of a rotating outer cup and stationary inner bob that to a good approximation

$$q(r_i) \sim \frac{2N\Omega}{1 - \left(\dfrac{r_i}{r_o}\right)^{-2N}} \quad , \tag{15}$$

where N is the exponent of the empirical power law fit of shear stress-shear rate data through

$$q = \kappa_1 \tau^N \qquad (N = \frac{1}{n}) \quad . \tag{16}$$

N would be determined from experimental results through

$$N = \frac{d \log \Omega}{d \log C} \quad . \tag{17}$$

It has been realised since the mid 1950's [81,82] that it is possible to determine normal stress differences in Couette flow by mounting pressure sensing transducers on the inner and outer cylinders. However, such instruments have never been developed commercially.

4.4 Cone-and-Plate (Figure 4)

A widely used instrument for measurement of shear flow rheological properties is the cone-and-plate system. This was first introduced by Mooney and Ewart [83] to correct for end effects in the coaxial cylinder

236

viscometer. This was called the "conicylindrical" viscometer and was applied to relatively low viscosity lattices. In 1945, Piper and Scott [84]described a biconical viscometer which was used in later years to characterise the viscosity of rubber [85,86]. Weissenberg [87] used this geometry as the basis of his "rheogoniometer" to measure both shear stresses and normal stresses in steady shear flow. This was first applied to polymer melts by Pollett and Cross [88] in 1950. The cone-and-plate instrument has been widely used since that time in many laboratories to measure both viscosity and normal stresses [9,41,45-48,51,89-93].

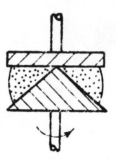

Steady
Shear

FIGURE 4.   Cone Plate Viscometer.

Generally, small cone angles are used in this instrument. In this geometry it is possible to develop a constant shear rate throughout which is given by

$$q = \Omega_1/\theta_0 , \tag{18}$$

where Ω_1 is the angular velocity of the moving member and θ_0 the angle between the cone and the plate.

Use of this instrument is generally limited to low shear rates. A flow instability in polymer melt occurs at about 1-10 sec^{-1} in which the melt emerges from the gap between the cone and the plate.

Measurements of the normal stresses as well as the shear stress described in this section are generally made using two types of commercial instruments. These are the Weissenberg Rheogoniometer originally developed by K. Weissenberg and currently manufactured by Sangamo Controls (Bognor Regis, United Kingdom) and the more recent Mechanical Spectrometer developed by J.M. Starita and C.W. Macosko and manufactured by Rheometrics (Union, New Jersey, USA). In recent years most investigations involving polymer melts have preferred the latter instrument.

4.5 Parallel Disc Viscometer (Figure 5)

It is also possible to measure viscosity in torsional flow between a stationary and a rotating disk. Instruments of this type have been reported in the literature for more than a century. The first instrument of this type to be used on bulk polymers was Mooney's 1934 'shearing disk' viscometer [94-98]. In this instrument a serrated disk is rotated in a sample fixed in a pressurized cavity. It was developed for application to rubber and its compounds and both pressurisation and serrations were meant to avoid slippage. This has become the standard quality control instrument of the rubber industry and we will return to it at a later point in this chapter. Beginning in the 1960's, various investigators [93,99,100] have used unpressurised smooth parallel disks to determine viscosity and normal stresses in shear flow of molten plastics.

Rheological properties in the parallel instrument are generally determined at the value of the shear rate at the outer radius of the disk which is

$$q_a = \frac{a\Omega}{H} .$$
 (19)

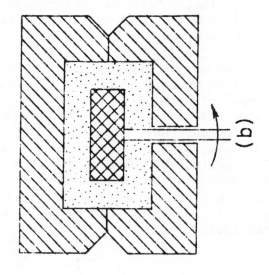

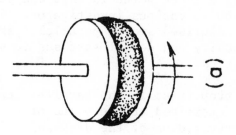

FIGURE 5 (a) Parallel Disc Viscometer.   (b) Mooney Shearing Disc Viscometer.

The shear stress and normal stress differences are given by

$$\tau \Big|_a = \frac{3C}{2\pi a^3} \left(1 + \frac{1}{3} \frac{d \ln C}{d \ln q_a} \right) ,\qquad (20a)$$

$$P_{11} - 2P_{22} + P_{33} \Big|_a = \frac{2F}{\pi a^2} \left[1 + \frac{1}{2} \frac{d \ln F}{d \ln q_a} \right] ,\qquad (20b)$$

where C is the torque and F the normal thrust.

As in the case of the cone-and-plate instrument, the simple parallel disk apparatus is limited to low shear rates. At higher shear rates, there is a flow instability in which the melt emerges from the gap between the disks.

The Sangamo Controls Weissenberg Rheogoniometer and the Rheometrics Mechanical Spectrometer can be placed in the parallel plate mode for making normal stress measurements of the type described above.

In Mooney's shearing disk viscometer, the shear stress at the outer radius of the disk may be expressed by:

$$\tau \Big|_a = \frac{C}{\pi a^3} f ,\qquad (21)$$

where f represents the contributions of the outer periphery. The magnitude of this contribution has been the concern of various investigators [94,97,101, 102], most recently and notably by Nakajima and Harrel [102] who find using a power law approximation

$$f = \frac{n+3}{4} \left[1 + \frac{\left[\left(\frac{2}{n} \right)^n (n+3) hH^n \right]}{2a^{n+1} \left[1 - \left(\frac{a}{a+\Delta} \right)^{2/n} \right]^n} \right]^{-1} ,\qquad (22)$$

240

where Δ is the distance between the outer disk radius
and cavity, h the thickness of the disk and n the
power law exponent

$$n \sim \frac{d \ln C}{d \ln q_a} .$$

(23)

4.6 Capillary Viscometer   (Figure 6)

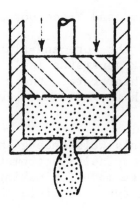

FIGURE 6. Capillary Viscometer.

The most widely used of all rheological instruments
for polymer melts is the capillary rheometer. It consists of a reservoir to which is connected a capillary
die with diameters generally in the range 0.05 cm to
0.20 cm. The melt is generally extruded at either a
constant volumetric throughput Q or at a constant
pressure. Capillary viscometric instruments have been
applied to Newtonian fluids for more than a century.
The first application of them to determine viscosity
characteristics of bulk polymer systems was probably
by Dillon and Johnston [103] in 1933. It became widely
used following the studies of Spencer and Dillon [104,
105] in 1948-49 because the experiments were not only

relatively simple to perform but the capillary rheometer reproduces an extrusion operation.

The shear stress along the wall of a capillary die
is closely related to the pressure drop Δp_{die} required
to extrude melt through the die. A force balance
gives

$$\tau_w = \frac{D \Delta p_{die}}{4L} \, , \tag{24}$$

where τ_w is the shear stress at the die wall. As
shown by Mooney and Black [96,106] and by Bagley [107],
large "ends" pressure losses are associated with the
process and one must in general write

$$P_T = 4 \tau_w \frac{L}{D} + \Delta p_{ends} \, . \tag{25}$$

The pressure loss Δp_{ends} is the sum of an entrance and
an exit pressure loss [65,108]

$$\Delta p_{ends} = \Delta p_{ent} + \Delta p_{exit} \, . \tag{26}$$

To determine τ_w, one must either obtain data at
several L/D ratios and determine it from the slope of
a p_T versus L/D plot or use pressure transducers along
the length of the die.

The shear rate at the die wall is given by the
Weissenberg equation

$$q_w = \frac{4Q}{\pi R^3} \left[\frac{3}{4} + \frac{1}{4} \frac{d \ln 4Q/\pi R^3}{d \ln \tau_w} \right] \tag{27}$$

and requires knowledge of both $4Q/\pi R^3$ and its slope
as a function of shear stress.

Generally, capillary viscometers may be used at
higher shear rate levels than rotational viscometers
for polymer melts. Flow instabilities variously
called simply extrudate distortion or melt fracture

represent the limitation at high extrusion rates [105, 109,110]. This flow instability generally occurs at die wall shear stress levels of about 10^5 Pa.

Many commercial capillary rheometers for viscosity measurements are available from companies in Europe, Japan and the U.S.A.

The capillary instrument has also been used to measure normal stresses although such methods have never gained general agreement. In polymer solutions, both the swell [111,112] and backward thrust force [113] of jets have been used. In polymer melts, the total ends pressure loss [114,115] and the exit pressure loss [65,108,116] have been applied. The efforts of Han in developing such an instrument should be noted. Of the techniques used for melts, the exit pressure loss technique has the more rigorous basis. It depends however on the presumption of fully developed flow existing up to the exit of the tube. This, however, appears not to be the case, especially in the very low Reynolds numbers flows that occur in polymer extrusion. Velocity field distortions at the exit of a tube have been determined for Newtonian fluids in slow flow by numerical solutions of the Navier-Stokes equations [117]. The velocity field variations in viscoelastic fluids should be at least qualitatively similar. However the magnitudes of the errors introduced are not known.

4.7 Slit Rheometer

The slit rheometer is similar in concept to the capillary apparatus only differing in the orifice cross-section. This type of instrument has been used for polymer melts by various investigators [65,116,118, 119] but has most notably been exploited by Han. Generally a series of flush mounted transducers are placed along the flow direction of the instrument. Pressure gradients are measured and may be converted to wall shear stress through the expression [65]

$$\tau_w = b \frac{dp_T}{dx} , \qquad (28)$$

where b is the half thickness of the channel. The wall shear rate is determined by the expression [54, 65,106]

$$q_w = \frac{3Q}{4ab^2} \left[\frac{2}{3} + \frac{1}{3} \frac{d \ln 3Q/4ab^2}{d \ln \tau_w} \right] \quad , \tag{29}$$

where a is the half width.

The slit rheometer operates in a similar region of high shear rates as the capillary instrument. It is generally limited at low shear rates by instrument sensitivity and at high shear rates by extrudate distortion.

Han and his coworkers have used exit pressure measurements to determine normal stresses with a slit rheometer. The procedure has not however gained general acceptance because of velocity-profile re-arrangements occurring near the exit, as discussed in the previous section. Again the magnitudes of the errors are not known, but the experimental data cited by Han suggest the problem might not be serious [92, 116] (see also Walters [54]).

5. ELONGATIONAL FLOW INSTRUMENTS

5.1 General

Many polymer processing operations involve uniaxial elongational flow rather than shear flow. The importance of such processes has led to the extensive development of instruments to carry out measurements of this type. The earliest study of instruments of this type were for viscous "Newtonian" liquids by Trouton [120] as early as 1906. The first studies on polymer melts were by Ballman [121], Cogswell [122], Meissner [123] and Vinogradov, Fikham and Radushkevich [124] in the late 1960's. Since that time many investigators have reported experimental studies of elongational viscosity of melts [45,46,125-131].

244

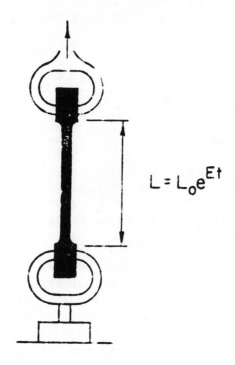

$$L = L_o e^{Et}$$

Ballman Method

FIGURE 7.(a) Elongational viscosity measurement methods
at constant elongation rate.

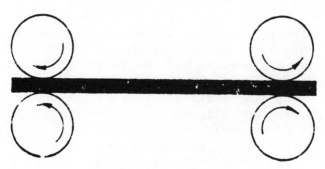

Meissner Method

FIGURE 7 (b)

Measurements of elongational viscosity are generally more difficult to carry out than shear flow studies. The problems involve the difficulty of maintaining a constant elongation rate or stress and achieving a steady state. All data reported to date on polymer melts has been on high molecular weight polyolefins, elastomers, styrenics, and their compounds.

5.2 Filament Stretching (Figure 7)

The major method of measurement of elongational viscosity is to stretch a filament. This may be done in a variety of ways. However, a general limitation exists in all of these experiments in that the fluid has to have a sufficient level of viscosity to be maintained in a controllable deformation field. This is the case for melts with viscosities of the order of 10,000 Pa.s. and greater, e.g. polyolefins and elastomers. However, it is not possible to carry out this type of deformation with lower viscosity melts such as polyesters and aliphatic polyamides which are often in the 100 Pa.s. range. It is also limited to deformation rates below 5 sec^{-1}.

There are alternate filament stretching methods of measurement of elongational viscosity. In the original method of Ballman [121] (see also [124,125]) a vertical thermostated filament is clamped at both ends and stretched at a rate dL/dt which varies in such a way as to maintain a constant rate of deformation k, i.e.

$$k = \frac{1}{L} \frac{dL}{dt} \quad ,$$

$$L = L(o) e^{Et} \quad ,$$

(30)

i.e. an exponential growth rate.

In the method of Meissner [123], a horizontal filament of length L is supported on the surface of a hot (thermostating) immiscible oil and held at both ends between pairs of toothed wheels with linear velocity V/2. The filament is drawn at both ends through the pair of wheels one of which contains a tension sensing instrument. The deformation rate k is

$$k = \frac{V/2}{L/2} = \frac{V}{L} \, . \tag{31}$$

Macosko and Lorntsen [127] and Ide and White [130] describe methods in which filaments are clamped at one end and taken up on a rotating roll which reduces the amount of filament being stretched as in the Meissner apparatus and attempts to produce a constant elongation rate. The latter instrument which floats a filament and then removes it onto a cold roll does a better job of this.

An instrument based on vertical extension of a ring specimen is described by Cotton and Thiele [132]. This instrument was designed to make measurements on elastomers which generally slip on rolls. Constant stress measurements of elongational viscosity [122, 126] have also been reported in the literature. A creep experiment using a constant force coupled with a cam to make it respond as a constant stress is utilised. The apparatus of Cogswell involves a filament being extended vertically on top of a bath, while that of Munstedt [126] involves a vertical sample immersed in a bath. The latter instrument allows only a limited extension determined by the apparatus design.

A more detailed review of instruments of this type is given by Dealy [133].

Recently, Rheometrics (Union, New Jersey, USA) has begun manufacture of the first commercially available instrument for measurements of an elongational viscosity. This is based on the apparatus of Munstedt [126].

5.3 Extrusion Method

The existence of elongational flow in the entry region of a capillary die was first noted by various investigators in the late 1960's [134, 135].

Cogswell [134] proposed and reported the use of pressure losses through orifice dies to determine elongational viscosity in polymer melts. The procedure has not gained any general acceptance because of the complex nature of the flow patterns in the die entrance regions and the extent of the approximation of converging entrance flow as an elongational flow with constant elongation rate. Early studies of the use of converging flow as elongational flow have been summarised by Cogswell [136] in 1978 where he suggested that the direction of development should be towards design of an instrument containing a die with a lubricated layer along the walls to prevent shearing motion and a position varying cross-section to ensure a constant elongation rate along the die axis. A publication in the same year by Everage and Ballman [131] notes that they have developed such a rheometer in the Monsanto Textile Company's Pensacola laboratories and report data for a polystyrene melt.

The two major disadvantages of the filament stretching method of measuring elongational viscosity is its limitations to very high viscosity melts and low extension rates. Such problems seem to be partly overcome in this apparatus. The controlled flow in the instrument allows much lower viscosity levels to be studied and Everage and Ballman indicate that they have studied systems with viscosities as low as 100 Pa.s. Indeed it would appear that the very reason for developing this instrument was to measure the elongational viscosity behaviour of low viscosity polymer melts such as polyesters and polyamides which are used to make synthetic fibres. Everage and Ballman also note that they can achieve elongation rates of 200 sec^{-1}. The elongation rate is limited by the occurrence of extrusion instabilities. A recent paper of Winter, Macosko and Bennett [137] also describes experiments of lubricated capillary flow to measure elongational viscosity.

Using a lubricated slit modification of this apparatus it is possible to make measurements of planar extensional viscosity. An apparatus of this type has been built and tested by C.W. Macosko in as yet unpublished research.

5.4 Inflation of Sheets

Existing methods of measurement of biaxial and planar extensional flow are based on the work of Denson [138-141] and involve inflation of clamped circular or high aspect-ratio rectangular sheets to form 'bubbles'. The material near the upper tip of the bubble undergoes a biaxial elongational flow for circular sheets and this can be programmed to be constant in rate if the movement of fiducial marks or the growth of the bubble is monitored and the inflation pressure varied to maintain this constant.

Early studies with this type of instrument were carried out at room temperature using elastomers such as polyisobutylene. More recent investigations [141, 144] have involved polymer melts.

A limitation of this apparatus involves the total strain that can be applied before the sheet becomes too distorted (beyond a hemisphere or half cylinder). Beyond this, one will not be able to develop a constant elongation rate flow. Thus one may often not be able to achieve a steady state elongational viscosity.

Another limitation is that the materials studied by such an apparatus must obviously have very high elongational viscosities. Denson's polymers often have viscosities in the range 10^8 to 10^9 Pa.s. Maerker and Schowalter [142] report results on systems with viscosities as low as 100,000 Pa.s.

6. SMALL STRAIN TRANSIENT EXPERIMENTS

6.1 General

An important class of rheological experiments performed on bulk polymer systems involves studying the response of these materials to small strains. The strain is sometimes applied as step functions and the relaxation of stress following its application is

measured. In other experiments a sinusoidal oscillatory deformation is applied and the stress response
to it is measured. Other variants have been investigated. Experiments of this type are best classified
together because their results are usually given a
uniform interpretation in terms of the theory of
linear viscoelasticity.

6.2 Stress Relaxation

An experiment which has been widely carried out on
elastomers [71,145-146] and other highly viscous
polymer systems [45] involves application of a strain
to a sample in extension or shear. The force response,
F(t), to the applied strain is measured. The data is
interpreted in terms of a relaxation modulus. For a
sample of initial cross-section A_o and length L_o which
is stretched in extension an amount ΔL, the relaxation
modulus E(t) is defined as

$$E(t) = \lim_{\Delta L \to o} \left[\frac{F(t)/A_o}{\Delta L/L_o} \right]. \tag{32a}$$

For a measurement carried out in shear, one uses a
sample of thickness H. We subject one of the surfaces
of the sample to a displacement L in a direction
perpendicular to its thickness 'H'. The shear modulus,
denoted by G(t) is given by:

$$G(t) = \lim_{\Delta L \to o} \left[\frac{F(t)/A}{\Delta L/H} \right]. \tag{32b}$$

Care needs to be applied in both sample preparation
and in being sure of developing and maintaining uniform strains. 'Residual' stresses often exist in
moulded tensile specimens. These can be removed by
dissolving samples and reprecipitating them as films
(e.g. on mercury) from which tensile specimens can be
cut.

250

Oscillatory
Shear

FIGURE 8 (a). Oscillatory Shear in a cone-plate instrument.

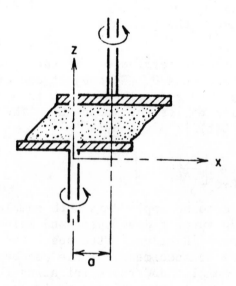

FIGURE 8 (b). Orthogonal Rheometer.

6.3 Sinusoidal Experiments (Figure 8)

A common experiment performed on polymer melts is
the determination of the stress response to an imposed
sinusoidal oscillation [44,45,52,53,143,145-147].
Measurements of this type on polymer solutions were
first carried out by Philippoff [147] in the 1930's
and became common on bulk polymers in the 1950's. The
results are usually interpreted in terms of the functions η', G', G", etc. (see Chapter 1).

Two different procedures are being widely used to
measure $G'(\omega)$ and $G''(\omega)$ on polymer melt systems. The
most widely used procedure for carrying out such
measurements has been to place the melt in the gap
between a cone and plate and oscillate one member
relative to the other with a shear strain [52,53,148-
150].

More recently a technique has been proposed by
Maxwell and Chartoff [151] which involves the rotation
of two parallel eccentric disks. It is shown that
fluid elements in this geometry undergo a periodic
sinusoidal deformation and the forces exerted on the
disk may be interpreted as $G'(\omega)$ and $G''(\omega)$. The
theory is described in some detail in the monograph
of Walters [54].

A survey of the literature and discussions with
academic industrial investigaters indicates that most
cone-and-plate sinusoidal oscillation measurements of
$G'(\omega)$ and $G''(\omega)$ are being carried out on the
Weissenberg Rheogoniometer and Rheometrics "Mechanical
Spectrometer". Eccentric disk measurements of these
properties have been carried out in some models of the
Rheometrics "Mechanical Spectrometer". For "linear"
viscoelastic materials, stress relaxation and dynamic
sinusoidal measurements lead to the same rheological
properties. This is discussed in Section 8.

7. FLOW BIREFRINGENCE MEASUREMENTS OF RHEOLOGICAL PROPERTIES

The determination of birefringence during flow is a

method of rheological property measurement unique to homogeneous systems of flexible polymer chains. The existence of birefringence in flowing polymer solutions has been known for a century [150]. However, it was only in the mid 1950's that Lodge [153] and Philippoff [154] showed the validity of the Rheo-Optical Law for polymer solutions. The first quantitative studies reported for polymer melts are those of Dexter, Miller and Philippoff [155] and of Wales and his coworkers [146,156-158]. More recent studies are reported by Han and Drexler [157], Vinogradov and his coworkers [160] and by Janeschitz-Kriegl and his team [161,162]. The studies of Janeschitz-Kriegl et al. involve stress transients and tests of the stress-optical law for such conditions.

The relationship of flow birefringent behaviour to stresses is determined by the Rheo-Optical Law which may be stated [153-162]

$$n_i - n_j = \bar{C}(p_i - p_j) \ . \tag{33}$$

Here n_i and n_j represent the principal refractive indices and p_i and p_j the principal stresses as developed during flow. (Here the principal stress represents the magnitudes of the principal axes of the ellipsoid of the stress tensor). $\bar{C}$ is the stress-optical constant which must be determined for each system studied. Eq. (33) is valid for flexible chain polymer melts such as polyolefin, polystyrene and elastomers. Great caution should be exerted when the presence of even small amounts of crystallinity are suspected [163].

In shear flow

$$p_1 - p_2 = [4p_{12}{}^2 + (p_{11} - p_{12})^2]^{\frac{1}{2}} \ , \tag{34a}$$

while in uniaxial extension

$$p_1 - p_2 = p_{11} - p_{22} \ . \tag{34b}$$

Measurements of birefringence on polymer solutions are traditionally carried out with a coaxial cylinder viscometer. However, measurements on polymer melts have, with few exceptions, been reported on parallel plate slit extrusion [157,159,160] or cone-and-plate devices [145,156,161] though Gortemaker et al. [162] have described a coaxial cylinder device. Dexter et al. [155] use Couette flow between parallel plates. Thus all of these instruments seek to use flow bire-fringence to measure shear flow properties. However, the technique may also be used for elongational flow. The author has seen a flow birefringence device mounted in a Meissner type elongational flow apparatus [123] by O. Ishizuka and K. Koyama of Yamagata University. Measurements in transient elongational flow such as melt spinning are common.

Birefringence is measured by investigating the trans-mission of a polarized monochromatic beam wavelength of light through the medium for a depth 'd'. Because the medium in the plane perpendicular to the trans-mitted wave has different refractive indices in differ-ent directions, the wave divides into two parts which take on differing velocities corresponding to the two refractive indices. The two waves become out of phase with each other. The beam exiting the apparatus goes through an analyzer which is a polarizer crossed at 90° with the initial direction of polarization.

The exiting wave is of form

$$\sin 2\alpha \sin R/2 \sin(pt - R/2) \ , \tag{35}$$

where R is the retardation

$$R = \frac{2\pi d \Delta n}{\lambda} \ . \tag{36}$$

α is the angle between the direction of polarization and n_1 or n_2, and λ is the wavelength. The birefring-ence Δn is determined either through the reduction in intensity of the exiting beam, the use of compensators, or through characterisation of the fringe patterns produced by the analyzer. The fringes represent

conditions where the contributions of the two waves to
the direction allowed passage by the analyzer cancel.

In the compensator method [158] one places a calibrated variable crystal thickness (perhaps by rotating
a crystal or translating a variable thickness) between
the apparatus and the analyzer and adjusts it until
it produces a retardation R which is then determined.
This gives Δn from Eq. (33). This is all that is
necessary in elongation flow as α of Eq. (35) is known,
but in shear flow one is unable to resolve $p_1 - p_2$
into $p_{11} - p_{22}$ and p_{12}. One needs the angle α (extin-
ction angle) between the nearest principal axes and
the direction of flow. It may be shown that

$$\tan 2\alpha = \frac{p_{11} - p_{12}}{2p_{12}} \ . \tag{37}$$

This may be obtained by rotating the polarizer and
analyzer relative to the apparatus until light is
extinguished.

In the total intensity method [155] one determines
the reduction in intensity of the transmitted light.
This is given by

$$\frac{I}{I_o} = \sin^2 2\alpha \ \sin^2 R/2 \ . \tag{38}$$

It is again necessary to measure α.

In non-homogeneous flows, it is necessary to determine Δn and α through locating and counting fringes.
Clearly, in a well known but variable stress field as
in the 12 plane in a slit, dark fringes appear when

$$2\alpha = N\pi \qquad \text{and} \qquad R/2 = N\pi \ ,$$

$$(N = 1, 2 \ \text{---}) \tag{39a,b}$$

where N is an integer. The fringes corresponding to
Eq. (39a) are called isoclinics and those of Eq. (39b)
are known as isochromatics. The latter correspond to

Δn values of

$$\Delta n = \frac{N\lambda}{\pi d} \, , \tag{40}$$

which gives the difference in principal stresses at that point. It is of course necessary to determine the order of the fringe. In order to separate iso-chromatics, which Eq. (40) applies to, from isoclinics, it is necessary to use a device such as quarter wave plates. These change linearly polarized to circularly polarized light and eliminate isoclinic fringes.

8. INTERPRETATION OF EXPERIMENTAL RESULTS IN TERMS OF PHENOMENOLOGICAL THEORIES

8.1 Linear Viscoelasticity [164]

Small strain transient experiments on polymer systems are generally interpreted in terms of the theory of linear viscoelasticity as developed by Boltzmann [164]. This formulation takes the 3-dimensional tensor form

$$p_{ij} = -p\delta_{ij} + 2\int_{o}^{\infty} G(z) \, e_{ij}^{(1)}(z) \, dz \, , \tag{41}$$

where p_{ij} is the covariant stress tensor, $e_{ij}^{(1)}$ the rate of deformation tensor, δ_{ij} the Kronecker delta and $G(t)$ the relaxation modulus function. The $G(t)$ function is exactly the function which is measured by shear stress relaxation.

It is possible to express other experimental quantities measured in small strain experiments in terms of $G(t)$. The tensile relaxation modulus $E(t)$ is related to $G(t)$ through

$$E(t) = 3G(t) \, . \tag{42}$$

The arguments relating $E(t)$ to $G(t)$ resemble those used to relate elongational to shear viscosity in a Newtonian fluid where a similar factor of three arises.

It is also possible to interpret $G'(\omega)$ and $G''(\omega)$ the storage and loss moduli in terms of Eq. (41). It is readily shown that

$$G'(\omega) = \omega \int_0^\infty G(s) \sin \omega s \, ds \, , \tag{43a}$$

$$\eta'(\omega) = \frac{G''(\omega)}{\omega} \int_0^\infty G(s) \cos \omega s \, ds \, . \tag{43b}$$

One generally interprets $G(t)$ in terms of a discrete or continuous spectrum of relaxation times:

$$G(t) = \sum_i G_i e^{-t/\lambda_i} \, , \tag{44a}$$

$$G(t) = \int_0^\infty H(\lambda) e^{-t/\lambda} \, d \ln \lambda \, . \tag{44b}$$

If we substitute Eqs. (44) into Eqs. (43), we obtain

$$G'(\omega) = \sum_i \frac{\omega^2 \lambda_i^2 G_i}{1 + \omega^2 \lambda_i^2} \, , \tag{45a}$$

$$G'(\omega) = \int_0^\infty \frac{\omega^2 \lambda^2 H(\lambda)}{1 + \omega^2 \lambda^2} \, d \ln \lambda \, , \tag{45b}$$

$$\eta'(\omega) = \frac{G''(\omega)}{\omega} = \sum_i \frac{G_i \lambda_i}{1 + \omega^2 \lambda_i^2} \, , \tag{46a}$$

$$\eta'(\omega) = \int_0^\infty \frac{H(\lambda) \, d\lambda}{1 + \omega^2 \lambda^2} \, . \tag{46b}$$

There are other results of interest in linear viscoelasticity. The zero shear rate asymptote of the viscosity is equal to the zero frequency asymptote

$$\lim_{\omega \to o} \eta'(\omega) = \lim_{q \to o} \eta = \eta_o$$

$$= \sum_i G_i \lambda_i = \int_0^\infty H(\lambda) d\lambda \quad . \tag{47}$$

Other linear viscoelastic results are of considerable interest. The steady state compliance J_e, the elastic recovery per unit stress following flow, is given by

$$J_e = \frac{\int_o^\infty \lambda H(\lambda) d\lambda}{\left[\int_o^\infty H(\lambda) d\lambda\right]^2} \quad . \tag{48}$$

8.2 Criminale - Erickson - Filbey (CEF) Equation

For long-duration steady-shearing flows, general non-linear viscoelastic fluid theories take the asymptotic form first given by Criminale, Ericksen and Filbey [165]:

$$P_{ij} = -pg_{ij} + 2\eta e_{ij}^{(1)} + 4N_2 e_{im}^{(1)} e_{mj}^{(1)}$$

$$- N_1 \left[\frac{\partial v_i}{\partial x_m} e_{mj}^{(1)} + e_{im}^{(1)} \frac{\partial v_m}{\partial x_j} \right] , \tag{49}$$

where N_1 and N_2 are normal-stress coefficients. This equation can be applied to velocity fields of the form

$$v_1 = v_1(x_2, x_3) , \quad v_2 = v_3 = 0 , \tag{50a}$$

which include flow in dies of varying cross-section.

It can also be applied to flows of the form

$$v_1 = v_1(x_2) \quad , \quad v_2 = 0 \quad , \quad v_3 = v_3(x_2) \quad , \tag{50b}$$

and in cylindrical coordinates

$$v_z = v_z(r) \quad , \quad v_\theta = r\omega(r) \quad , \quad v_r = 0 \; . \tag{50c}$$

This includes combined Couette and pressure flow and helical flow.

From the CEF theory it is possible to argue more than intuitively that for viscoelastic fluids all viscometric shear flows yield the same rheological response. It also suggests in what kind of problem (helical flow, conduit flow) these properties may be used in theoretical predictions of flow behaviour.

9. POLYMER STRUCTURE AND RHEOLOGICAL PROPERTIES

There have been extensive studies of the influence of structure on the rheological properties of polymer melts. The results of these investigations are used by industrial scientists to interpret the processing characteristics of polymer melts. It is thus worthwhile to review these results.

Much of the effort on studies of rheological properties of polymer melts has been devoted to the relaxation modulus G(t) or the spectrum of relaxation times $H(\lambda)$. Investigations of $H(\lambda)$ date to the work of Andrews, Hofman-Bang and Tobolsky [145] in the late 1940's. Generally, using stress relaxation, Tobolsky [146] in his 1960 monograph summarized his fifteen years of investigations of $H(\lambda)$. More recent experimental studies have been reported by Masuda and Onogi [166,167] amongst others, frequently using sinusoidal oscillation studies. The generally accepted view is that in narrow molecular weight distribution systems, there is a wedge-box distribution

Wedge $H(\lambda) = \dfrac{A}{\lambda}$ $\quad \lambda < \lambda_1$ $\quad\quad$,

box $\quad\quad H(\lambda) = H_o$ $\quad \lambda_2 < \lambda < \lambda_m$.

$$(51)$$

Generally, A, H_o, λ_1 and λ_2 depend only on the polymer backbone. λ_m is a strong function of molecular weight, varying according to

$$\lambda_m = K_{\lambda m} M^{3.5} . \qquad (52)$$

With broad distribution samples, the box spreads out to larger times λ. Some of the data however of Masuda et al. suggest that for very narrow distributions there may be a spike rather than a box.

It is of considerable interest to study other linear viscoelastic properties which are integrals of $H(\lambda)$ though the particular properties are often measured in different manners. Most important is the zero shear viscosity η_o which in linear monodisperse systems shows a similar dependence to λ_m. In polydisperse systems [9,168,169]

$$\eta_o = K_{\eta_o} M_w^{3.5} . \qquad (53)$$

For monodisperse branched systems [170], η_o is lower than the prediction of Eq. (53) at low deformation rates but increases rapidly at molecular weights above 100,000. At high molecular weights η_o is much higher for branched than for linear polymers.

An important linear viscoelastic characteristic is the steady state compliance J_e defined by Eq. (54). Many investigations have studied the influence of molecular weight and its distribution on J_e. Generally it is found that J_e is relatively independent of molecular weight but strongly dependent on its distribution. Broad distributions give very large J_e. Empirical equations developed have been of the form [9,171,172]

$$J_e = K_{J_e} \left(\frac{M_z}{M_w}\right)^{3.5} . \tag{54}$$

Linear viscoelastic properties of flexible polymer chains exhibit very striking temperature dependence. $G(t)$, $G'(\omega)$ etc. data taken at different temperatures may be shifted together to form 'master curves'. These shift properties were first used extensively by Tobolsky and his colleagues during the 1940's. They are generally expressed as [143]:

$$G(t,T) = \frac{\rho T}{\rho_o T_o} G(t/a_T, T_o) . \tag{55}$$

The shift factor a_T depends upon the value $T - T_g$ where T_g is the glass transition temperature.

The viscosity-shear rate behaviour is also dependent upon molecular weight distribution. This has been the subject of extensive study. It has become established following Vinogradov and Malkin [173] that plots of η/η_o versus η_o/q are independent of temperature but strongly depend upon molecular weight distribution. The curves fall off more rapidly as the molecular weight distribution broadens [9]. Similar methods of correlation of shear viscosity data have been reported by other research teams [174,175] frequently in terms of η/η_o vs a dimensionless λq plot where λ is a characteristic time of the fluid. Such plots also indicate a more rapid decrease of η/η_o for broad distribution samples.

The principal normal stress difference exhibits a strong dependence on molecular weight distribution. This might be predicted from the theory of second-order fluids [176] which represents the asymptotic slow-flow behaviour. Specifically this predicts

$$P_{11} - P_{22} = 2J_e P_{12}^2 = 2J_e \eta_o^2 q^2 . \tag{56}$$

From Eqs. (53) and (54) one would surmise

$$P_{11} - P_{22} \simeq \left(\frac{M_z}{M_w}\right)^{3.5} \quad , \quad P_{12}^2 \sim M_z^{3.5} M_w^{3.5} q^2 . \qquad (57)$$

White and Kondo [177] observe that the principal normal stress difference coefficient at 1 sec^{-1} in a series of polystyrenes varies with about the 6th power of molecular weight. Minoshima, White and Spruiell [9] report a similar variation with polypropylenes. Oda, White and Clark [178] develop a general correlation for $p_{11} - p_{22}$ in polystyrenes by plotting it versus p_{12}. The plots are independent of temperature and average molecular weight but vary strongly with molecular weight distribution – the broader the distribution the larger $p_{11} - p_{22}$. A similar correlation is found for polypropylenes by Minoshima et al. [9].

Studies of the influence of molecular weight distribution on elongational viscosity are presented by Minoshima, White and Spruiell [9]. They find that broadening the distribution increases filament instability and leads to necking and ductile failure. They indicate that narrower distribution linear samples ($M_w/M_n \sim 2.5-5$) show an increasing elongational viscosity function. Broader distribution functions exhibit a decreasing function.

10. QUALITY CONTROL MEASUREMENTS

10.1 General

In this section we turn our attention to quality control instrumentation used in the polymer industry. Specifically we shall be concerned with instruments which are used by polymer manufacturers and often by fabrications to specify and control the polymer either sold to customers or to determine whether it should be used in production in a specific process. There are two instruments of this type whose breadth of use is very great and dominates all other procedures used in the trade. These are the 'Mooney viscometer' used in the rubber industry and the 'melt indexer' used in the plastics industry. We shall discuss each of these in turn. The above instruments yield rapid single point measurements of quantities related to shear viscosity. It has long been realised that such procedures give only rough estimates of the rheological behaviour of elastomers and polymer melts and are inadequate especially as production rates approach conditions where flow instabilities occur or where it is necessary to closely control some aspect of a product such as extrudate dimensions. Various 'multiple point' modifications of the Mooney viscometer and melt indexing have been proposed for this purpose.

10.2 Mooney Viscometer [94-98,101-102] (Figure 5b)

We have already discussed the Mooney viscometer in Section 4.5. It consists of a rotating disk in a cylindrical cavity. By the specifications of the American Society for Testing Materials (ASTM), the diameter of the disk is 19.05 mm and has thickness 5.54 mm. A single point viscosity measurement is made at a rotor speed of two r.p.m. after a period of four minutes. The measurements are reported in terms of a calibrated torque measurement on a standard dial and referred to as the ML-4 reading. Here 'M' refers to the Mooney viscometer, L to a large disk size (distinguished from a smaller diameter disk used for vulcanization 'scorch' tests) and 4 for the four minutes of test. Measurements are carried out at 100°C. Typical values of ML-4 for commercial synthetic rubbers are in the range 45-55.

It is possible to convert the ML-4 readings directly to torque, C, shear stress, τ, and viscosity values. The shear stress at the outer edge of the disk is given by Nakajima and Harrel [102] as

$$C = 8.30 \quad 10^{-2} \ (DR) \ (Newton/meter) \ ,$$

$$\tau_a = 0.382 \ (DR) \cdot f \times 10^4 \ (pascal) \ , \tag{58}$$

where DR is the Mooney unit dial reading and f is the function of Eqs. (21) and (22) which is of order 0.55.

An ML-4 of 50 corresponds to a shear stress of 10^5 Pa. in the torsional flow at the outer edge of the disk. The shear rate at this position is 1.5 sec^{-1}. This corresponds to a viscosity of 75,000 Pa.s. (750,000 poise).

The Mooney viscometer was devised in 1934 by Melvin Mooney [94] and appears to have been adopted by the U.S. Rubber Company as a quality control instrument. In 1942, the Technical Committee of the Rubber Reserve Company, the U.S. Government organisation which ran the synthetic rubber programme during World War II, recommended that the Mooney shearing disk viscometer should be the standard instrument for controlling the processing of the GR-S synthetic rubber. The National Bureau of Standards determined an optimum instrument design and testing procedure [95].

10.3 Melt Indexer [179] (Figure 6)

The melt indexer developed by DuPont is the standard quality control device used in the plastics industry. Polyethylenes, polypropylene and polystyrene are generally specified by melt index (MI). It is basically an extrusion rheometer of the type described in Section 4.6. This apparatus uses a single capillary of diameter 0.655 mm and length/diameter ratio of 12 and extrusion is carried out due to a pressure applied through a dead weight. The melt index is the weight in grams extruded for a ten minute test.

The ASTM specifies a series of testing conditions A-K which involves changes in temperature (125°-275°C) and applied dead load (325-21,600 grams) or pressures (0.5-29.5 atmospheres). The specifications are chosen to give melt indices between 0.15 and 25 (grams per 10 minutes). Condition E is most widely used. This involves a temperature of 190°C and a pressure of 2.95 atmospheres.

The melt index is related to the reciprocal of the viscosity. However, it is not an exact correspondence because there is no correction for the ends pressure loss or the Weissenberg-Rabinowitsch shear rate correlation.

Junk, Wortberg and Menges [180] have suggested using
an adaptation of equation (27) to relate MI to q_w and
then applying a mean Vinogradov-Malkin [173] universal
viscosity curve to obtain the full viscosity - shear
relationship. Such a procedure is probably a reasonable first approximation but should be applied with
caution.

10.4 Other Off-Line Instruments

10.4.1 General Comments

It is generally agreed among industrial polymer
rheologists that melt index or Mooney viscosity
measurements are inadequate to specify processing
characteristics. They are especially inadequate in
representing variations in elasticity and elongational
flow behaviour found among different commercial
nominally-equivalent polymers. This has led to a
range of different parameters and instruments being
used by industrial rheologists to supply the missing
parameter. In some cases these additional measurements may be done in day-to-day quality control testing and supplied to interested customers perhaps
including fabricating divisions of the same company.
In other cases such additional measurements may only
be carried out as a part of trouble-shooting particular
problems. It would also appear that in some instances,
quality control problems arising from insufficient
specification from melt index and Mooney viscosity
lead immediately to very basic rheological measurements
of the type described earlier. This has much to say
for it, but it can be time consuming.

10.4.2 Extrudate Swell (Extrudate Shrinkage)(Figure 1c)

A very common method of control testing of elastic
recovery in polymer melts and compounds is to measure
the swell of extrudates. Studies using various procedures of swell measurement have been reported in the
literature though none has been standardised [45,48,
103-105,181-184]. White and Roman [183] compare a

range of methods of extrudate swell measurements on
polyolefins and polystyrene ((i) cold frozen swell,
(ii) annealed extrudates, (iii) photographed extrudates emerging into air, (iv) photographed extrudates
emerging into isothermal chambers of temperature
equal to die) and noted the differences in values.
This was attributed variously to delayed elastic
recovery encountered in the second and fourth measurements and variations in density between the first and
second (data taken at room temperature) and the third
and fourth (data at 180°C).

Extrudate swell has traditionally been measured as
indicated by photographs or on extrudates with either
a micrometer or by determining weight per unit length
(usually performed on soft elastomers). Recently the
Monsanto Industrial Chemicals Co. (Akron, Ohio) has
introduced an extrudate swell detector based on interaction with a scanning laser beam placed just below
the capillary die exit. This allows for rapid, almost
instantaneous, swell measurements.

10.4.3 Rheotens [185]

The Rheotens is the only commercial quality control
instrument which gives a measure of the elongational
flow properties of polymer melts. It is widely used
in Europe but not in the U.S.A. or Japan. This
apparatus, which was developed by Meissner [185],
consists of extruding a stream vertically downwards
from a capillary rheometer which is then drawn down
between two rotating gears. The gears are connected
to a variable speed motor and a tension sensing device.
Generally, the extrudate is drawn down at an ever
increasing programmed rate and both the tension trace
versus time or drawdown and the critical drawdown
ratio and tensile strength are measured. This apparatus is made by Goettfert (Buchen-Odenwald, West
Germany).

The Rheotens tension is related to the elongational
viscosity but the situation is complex. The kinematics are not consistent with a constant deformation
rate and the period of drawdown is very short.

Furthermore, the experiment is highly non-isothermal.

10.5 On-Line Quality Control Measurements

In this section we discuss quality control measurements which involve in-line placement of instruments rather than testing of samples removed from the processing apparatus. Most apparatus involves two pressure taps along the direction of flow in a conduit. At a fixed flow rate the pressure drop gives the conduit pressure gradient which is proportional to the viscosity.

Some modifications of this approach have been proposed and one successfully commercialised in recent years. C.D. Han and Seiscor (Division of Seismograph Service, Tulsa, Oklahoma) have introduced the use of exit pressure loss measurements in addition to constant cross-section pressure gradients. Melt is bled through a side-stream to give such measurements. The pressure gradient yields a shear viscosity. There is an extensive enough literature on melts in this type of apparatus [65,108,116,119] to indicate that the exit pressure loss gives a reliable representation of melt elasticity which can be correlated with normal stresses.

This area remains in flux and other designs have been proposed. An apparatus developed by A.S. Lodge with Seiscor uses a series of three pressure taps containing differently mounted transducers. Two transducers along the line of flow are similarly mounted (flushed or recessed) and give the pressure gradient and shear viscosity. Two transducers at a particular cross-section, through comparison of flushed and recessed pressure loss measurements, yield an approximate value of the normal stresses. As yet there appear to be no experimental verifications of this apparatus for polymer melts though it has been used successfully with polymer solutions.

11. DETAILED EXPERIMENTAL INVESTIGATIONS OF FLOW IN THE POLYMER MELT PROCESSING OPERATIONS

In many instances melt flow processing problems are resolved by direct observation of the flow or stress patterns using transparent fixtures. Procedures of this type were first introduced by Spencer and Gilmore [186] and Tordella [187] in conjunction with investigations of the mechanisms of injection moulding and extrusion. Spencer and Gilmore observed moving melt fronts and distinguished jetting from front motion mould filling. Tordella [187] introduced the use of markers to study flow patterns in polymer melts in the entrance region of a die and was the first to observe vortices in low density polyethylene in this geometry. Later he was the first to make on-line flow birefringence studies of polymer processing operations [188]. These efforts have been followed up by various investigations mainly to study extrusion into dies [177,189, 190] and injection moulding [191,192] including foaming [193,194] and two phase [195,196] polymer melts.

Most of the above measurements are qualitative in character. It is often useful and important to have quantitative measurements. Adams, Whitehead and Bogue [197] initiated quantitative measurements of stress fields in flowing polymer fluids using birefringence measurements in the tradition of photoelasticity. Extensions of such procedures to polymer melts have been few. The noteworthy study of Han and Drexler [159] in converging flow and of Han [198] for complex dies should be cited. The procedures of these authors use the methods of Section 7 to determine stress fields.

Little appears to have been done to quantitatively measure velocity fields in polymer melts. However, with the development of methods such as Laser-Doppler anemometry, progress should be expected.

REFERENCES : CHAPTER 5

[1]   P.J. Flory, "Principles of Polymer Chemistry",
      Chapters II, III, VI, Cornell University Press,
      Ithaca, 1953

[2]   J.M. Schultz, "Polymer Materials Science",
      Chapters 1, 5, Prentice Hall, Englewood Cliffs,
      1974

[3]   Reference 1, Chapters VII, VIII

[4]   Reference 2, Chapter 6

[5]   C. Tanford, "Physical Chemistry of Macromolecules",
      Wiley, New York, 1960

[6]   M.J.R. Cantow, Editor, "Polymer Fractionation",
      Academic Press, New York, 1967

[7]   H. Morawetz, "Macromolecules in Solution",
      2nd ed., Wiley, New York, 1975

[8]   T. Kotaka and J.L. White, Macromolecules 7,
      (1974) 105

[9]   W. Minoshima, J.L. White and J.E. Spruiell,
      Polym. Eng. Sci., (in press)

[10] A. Sharples, "Introduction to Polymer Crystal-
      lization", Arnold, London, 1966

[11] L.R.G. Treloar, "Physics of Rubber Elasticity",
      2nd ed., Chapters I, XI, Oxford, 1958

[12] J.E. Spruiell and J.L. White, Polym. Eng. Sci.,
      15 (1975) 660

[13] R.F. Boyer, Rubber Chem. Technol., 36 (1963) 1303

[14] V.G. Bankar, J.E. Spruiell and J.L. White,
      J. Appl. Polym. Sci., 21 (1977) 2348

[15] A.N. Gent, J. Polym. Sci., A3, 1965, 3787;
      ibid, 4 (1966) 447

[16] J. Shimizu, K. Toriumi and K. Tamai, Sen-i-Gakkaishi, 33 (1977) T-208

[17] G. Pezzin and G.B. Gechele, J. Appl. Polym. Sci., 8 (1964) 2195

[18] J.C. Staten, J.P. Keller, R.C. Kowalski and J.W. Harrison, U.S. Patent 3,551,943 (1971)

[19] R.C. Kowalski, U.S. Patent 3,563,972 (1971)

[20] T. Hancock, English Patent 7344 (1837)

[21] T. Hancock, "Origin and Process of the Caoutchouc or India Rubber Manufacture", Longman, London, 1857

[22] W.F. Busse, Ind. Eng. Chem., 24 (1932) 140

[23] W.F. Busse and E.N. Cunningham, Proc. Rubber Technol. Conf., (1938) 288

[24] M. Pike and W.F. Watson, J. Polym. Sci., 9 (1952) 229

[25] W.F. Watson, Trans. I.R.I., 29 (1953) 32

[26] G. Ayrey, C.G. Moore and W.F. Watson, J. Polym. Sci., 19 (1956) 1

[27] P. Goodman and A.B. Bestul, J. Polym. Sci., 18 (1955) 235

[28] R.J. Ceresa and W.F. Watson, J. Appl. Polym. Sci., 1 (1959) 101

[29] G. Kraus and K.W. Rollman, J. Appl. Polym. Sci., 8 (1964) 2585

[30] R.S. Porter and J.F. Johnson, J. Appl. Phys., 35 (1964) 3149

[31] H. Homma, N. Tagata and H. Hibino, Nippon Gommu Kyokaishi, 41 (1968) 242

270

[32] D.J. Angier and W.F. Watson, J. Polym. Sci., 18 (1955) 129, Trans. I.R.I. 33 (1957) 22

[33] D.J. Angier and W.F. Watson, J. Polym. Sci., 20 (1956) 235; ibid, 25 (1957) 1

[34] D.J. Angier, E.D. Farlie and W.F. Watson, Trans. I.R.I., 34 (1958) 8

[35] D.J. Angier, R.J. Ceresa and W.F. Watson, J.Polym. Sci., 34 (1959) 699

[36] R.J. Ceresa, Polymer, 1 (1960) 72, 397, 477, 488

[37] J.L. Amos, Polym. Eng. Sci., 14 (1974) 1

[38] S.L. Rosen, Ann. N.Y. Acad. Sci., 35 (1973) 480

[39] M. Morton, Editor, "Rubber Technology" 2nd ed., van Nostrand Reinhold, New York, 1973

[40] W.V. Titow and B.J. Lanham, "Reinforced Thermo-plastics", Applied Science Publishers Ltd., London, 1975

[41] B.L. Lee and J.L. White, Trans. Soc. Rheology, 19 (1975) 481

[42] T.I. Ablazova, M.V. Tsebenko, A.B.V. Yudin, G.V. Vinogradov and B.V. Yarlykov, J. Appl. Polym. Sci., 19 (1975) 1781

[43] K. Iwakura and T. Fujimura, J. Appl. Polym. Sci., 19 (1975) 427

[44] G.V. Vinogradov, A. Ya Malkin, E.P. Plotnikova, O. Yu Sabsai and N.E. Nikolayeva, Int. J. Polym. Mat., 2 (1972) 1

[45] V.M. Lobe and J.L. White, Polym. Eng. Sci., 19 (1979) 617

[46] H. Tanaka and J.L. White, Polym. Eng. Sci., (in press)

[47] F.M. Chapman and T.S. Lee, SPE J., 26 (Jan 1970) 37

[48] N. Minagawa and J.L. White, J. Appl. Polym. Sci., 20 (1976) 501

[49] J.E. O'Connor, Rubber Chem. Technol., 50 (1977) 445

[50] L. Czarnecki and J.L. White, J. Appl. Polym. Sci., (in press)

[51] T.S. Lee, Proc. 5th Int. Rheol. Congr., 4 (1970) 421

[52] H. Munstedt, Proc. 7th Int. Rheol. Congr., (1976) 496

[53] Y. Aoki, J. Soc. Rheol. Japan, 7 (1979) 20

[54] K. Walters, "Rheometry", Chapman and Hall, London, 1975

[55] R.B. Bird, W.E. Stewart and E.N. Lightfoot, "Transport Phenomena", Wiley, New York, 1960

[56] H.C. Brinkman, Appl. Sci. Res., A2 (1951) 120

[57] R.B. Bird, SPE J. (1955) 35

[58] R. Turian, Chem. Eng. Sci., 20 (1965) 771

[59] H.W. Cox and C.W. Macosko, AIChE J., 20 (1974) 785

[60] E.C. Bernhardt, Editor, "Processing of Thermoplastic Materials", Reinhold, New York, 1959

[61] J.M. McKelvey, "Polymer Processing", Wiley, New York, 1962

[62] J.L. White, Rubber Chem. Technol., 42 (1969) 257

[63] Z. Tadmor and I. Klein, "Engineering Principles of Plasticating Extrusion", van Nostrand Reinhold, New York, 1970

[64] A. Ziabicki, "Fundamentals of Fibre Formation", Wiley, New York, 1976

[65] C.D. Han, "Rheology in Polymer Processing", Academic Press, New York, 1976

[66] S. Middleman, "Fundamentals of Polymer Processing", McGraw-Hill, New York, 1977

[67] Z. Tadmor and I. Klein, "Engineering Principles of Plasticating Extrusion", van Nostrand Reinhold, New York, 1970

[68] N.V. Zakharenko, F.S. Tolstukhina and G.M. Bartenev, Rubber Chem. Technol., 35 (1962) 236

[69] S. Middleman, Trans. Soc. Rheol. 13 (1969) 123

[70] C. Goldstein, Trans. Soc. Rheol., 18 (1974) 357

[71] I. Furuta, V.M. Lobe and J.L. White, J. Non-Newt. Fluid Mech., 1 (1976) 207

[72] M. Mooney, Physics, 7 (1936) 413

[73] W. Philippoff and F.H. Gaskins, J. Polym. Sci., 21 (1956) 205

[74] F.N. Cogswell, Plastics and Polymers, 43 (1973) 39

[75] K. Weissenberg, Nature, 159 (1947) 310

[76] F.D. Farrow, G.M. Lowe and S.M. Neale, J. Textiles Inst., 19 (1923) T18

[77] M. Mooney, J. Rheology, 2 (1931) 231

[78] I.M. Krieger and H. Elrod, J. Appl. Phys., 23 (1952) 147

[79] I.M. Krieger and H. Elrod, J. Appl. Phys., 24 (1953) 134

[80] I.M. Krieger, Trans. Soc. Rheol., 12 (1968) 5

[81] H. Markovitz, Trans. Soc. Rheol., 1 (1957) 37

[82] H. Markovitz, Polymer Letters, 3 (1965) 3

[83] M. Mooney and R.H. Ewart, Physics, 5 (1934) 350

[84] G.H. Piper and J.R. Scott, J. Sci. Inst., 22 (1945) 206

[85] D.H. Saunders and L.R.G. Treloar, Trans. I.R.I., 24 (1948) 92

[86] L.R.G. Treloar, Trans. I.R.I., 25 (1949) 167

[87] K. Weissenberg, Proc. 1st Int. Rheol. Cong., (1948)

[88] W.F.O. Pollett and A.H. Cross, J. Sci. Inst., 27 (1950) 209

[89] R.G. King, Rheol. Acta, 5 (1966) 35

[90] T.F. Ballenger, I.J. Chen, J.W. Crowder, G.E. Hagler, D.C. Bogue and J.L. White, Trans. Soc. Rheol., 15 (1971) 195

[91] J. Meissner, J. Appl. Polym. Sci., 16 (1972) 2877

[92] C.D. Han, K.U. Kim, N. Siskovic and C.R. Huang, J. Appl. Polym. Sci., 17 (1973) 95

[93] B.L. Lee and J.L. White, Trans. Soc. Rheol., 18 (1974) 467

[94] M. Mooney, Ind. Eng. Chem. Anal. Ed., 6 (1934) 147

[95] R. Taylor, J.H. Fielding and M. Mooney, Symposium on Rubber Testing, p. 36 ASTM (1947)

[96]   M. Mooney, J. Colloid Sci., 2 (1947) 69

[97]   M. Mooney, Proc. Int. Rubber Conf. (Washington) (1959) 368

[98]   M. Mooney, Rubber Chem. Technol., 35 (5) XXVII (1962)

[99]   K. Sakamoto, N. Ishida and Y. Fukusawa, J. Polym. Sci., A-2, 6 (1968) 1999

[100] L.L. Blyler, Trans. Soc. Rheology, 13 (1969) 39

[101] J.L. White and N. Tokita, J. Appl. Polym. Sci., 9 (1965) 1929

[102] N. Nakajima and E.R. Harrell, Rubber Chem. Technol., 52 (1979) 9

[103] J.H. Dillon and N. Johnston, Physics, 4 (1933) 225

[104] R.S. Spencer and R.E. Dillon, J. Colloid Sci., 3 (1948) 163

[105] R.S. Spencer and R.E. Dillon, J. Colloid Sci., 4 (1949) 241

[106] M. Mooney and S.A. Black, J. Colloid Sci., 7 (1952) 204

[107] E.B. Bagley, J. Appl. Phys., 28 (1957) 624

[108] C.D. Han, M. Charles and W. Philippoff, Trans. Soc. Rheology, 13 (1969) 455

[109] J.P. Tordella, J. Appl. Phys., 27 (1956) 454

[110] J.L. White, Appl. Polym. Symp., 20 (1973) 155

[111] F.H. Gaskins and W. Philippoff, Trans. Soc. Rheology, 3 (1959) 181

[112] A.B. Metzner, W.T. Houghton, R.A. Sailor and
       J.L. White, Trans. Soc. Rheology, 5 (1961)
       133

[113] C.R. Shertzer and A.B. Metzner, Proc. 4th Int.
       Rheol. Cong., 2 (1965) 603

[114] W. Philippoff and F.H. Gaskins, Trans. Soc.
       Rheology, 2 (1958) 263

[115] J.L. White and A. Kondo, J. Appl. Polym. Sci.,
       21 (1977) 2289

[116] C.D. Han, Trans. Soc. Rheology, 18 (1974) 103

[117] R.I. Tanner, Appl. Polym. Symp., 20 (1973) 201

[118] J.L.S. Wales, J.L. den Otter and H. Janeschitz-
       Kriegl, Rheol. Acta, 4 (1965) 146

[119] C.D. Han, J. Appl. Polym. Sci., 15 (1971)
       2567, 2579, 2591

[120] F. Trouton, Proc. Roy. Soc., A77 (1906) 426

[121] R.L. Ballman, Rheol. Acta, 4 (1965) 137

[122] F.N. Cogswell, Plastics and Polymers, 36 (1968)
       109

[123] J. Meissner, Rheol. Acta, 8 (1969) 78

[124] G.V. Vinogradov, B.V. Radushkevich and V.D.
       Fikham, J. Polym. Sci., A-2, 8 (1970) 1

[125] J.F. Stevenson, AIChE J., 18 (1972) 540

[126] H. Munstedt, Rheol. Acta, 14 (1975) 1077

[127] C.W. Macosko and J.M. Lorntsen, SPE Antec
       Tech. Papers (1973) 461

[128] A.E. Everage and R.L. Ballman, J. Appl. Polym.
       Sci., 20 (1976) 1137

[129] H.M. Laun and H. Munstedt, Rheol. Acta, 17 (1978) 415

[130] Y. Ide and J.L. White, J. Appl. Polym. Sci., 22 (1978) 1061

[131] A.E. Everage and R.L. Ballman, Nature, 273 (1978) 213

[132] G.R. Cotten and J.L. Thiele, Rubber Chem. Technol., 57 (1979) 749

[133] J.M. Dealy, J. Non Newt. Fluid Mech., 4 (1978) 9

[134] F.N. Cogswell, Rheol. Acta, 8 (1969) 187

[135] A.B. Metzner, E.A. Uebler and C.F. Chan Man Fong, AIChE J., 15 (1969) 750

[136] F.N. Cogswell, J. Non Newt. Fluid Mech., 4 (1978) 9

[137] H.H. Winter, C.W. Macosko and K.E. Benner, Rheol. Acta, 18 (1979) 323

[138] C.D. Denson and R.J. Gallo, Polym. Eng. Sci., 11 (1971) 174

[139] D.D. Joye, G.W. Poehlein and C.D. Denson, Trans. Soc. Rheol., 16 (1972) 421

[140] C.D. Denson and D.L. Crady, J. Appl. Polym. Sci., 18 (1974) 1611

[141] C.D. Denson and D. Hylton, Proc. 7th Int. Rheol. Cong., (1976) 386

[142] J.M. Maerker and W.R. Schowalter, Rheol. Acta, 13 (1974) 627

[143] E.D. Baily, Trans. Soc. Rheology, 18 (1974) 635

[144] J.M. Dealy, Paper presented at AIChE Meeting, Miami Beach, November 1978

[145] R.D. Andrews, N. Hofman-Bang and A.V. Tobolsky, J. Polym. Sci., 3 (1948) 669

[146] A.V.Tobolsky, "Structure and Properties of Polymers", Wiley, New York, 1960

[147] W. Philippoff, Phys. Zeit, 35 (1934) 884

[148] J.W.C. Adamse, H. Janeschitz-Kriegl, J.L. Den Otter and J.L.S. Wales, J. Polym. Sci. A-2, 6 (1968) 871

[149] N.J. Mills and A. Nevin, J. Polym. Sci., 9 (1971) 267

[150] I.J. Chen and D.C. Bogue, Trans. Soc. Rheol., 16 (1972) 59

[151] B. Maxwell and R. Chartoff, Trans. Soc. Rheol., 9 (1965) 41

[152] J.C. Maxwell, Proc. Roy. Soc., 22 (1873) 46

[153] A.S. Lodge, Trans. Faraday Soc., 52 (1956) 120

[154] W. Philippoff, J. Appl. Phys., 27 (1956) 984

[155] F.D. Dexter, J.C. Miller and W. Philippoff, Trans. Soc. Rheology, 5 (1961) 193

[156] J.L.S. Wales and H. Janeschitz-Kriegl, J. Polym. Sci., 5 (1967) 781

[157] J.L.S. Wales, Rheol. Acta, 8 (1969) 38

[158] J.L.S. Wales and W. Philippoff, Rheol. Acta, 12 (1973) 25

[159] C.D. Han and L.H. Drexler, J. Appl. Polym. Sci., 17 (1973) 3429

278

[160] V.I. Brizitsky, G.V. Vinogradov, A.I. Isaev and
Yu Ya Podolsky, J. Appl. Polym. Sci., 20
(1976) 25; and ibid, 22 (1978) 665

[161] F.H. Gortemaker, M.G. Hanson, B. de Cindio,
H.M. Laun and H. Janeschitz-Kriegl, Rheol.
Acta, 15 (1976) 256

[162] F.H. Gortemaker, H. Janeschitz-Kriegl and
K. te Nijenhuis, Rheol. Acta, 15 (1976) 487

[163] L.R.G. Treloar, Trans. Faraday Soc., 36 (1940)
538; ibid, 37 (1941) 84; ibid, 43 (1947) 284

[164] L. Boltzmann, Sitzungber Akad Wiss Wien, 70
(1874) 275

[165] W.O. Criminale, J.L. Ericksen and G.L. Filbey,
Arch. Rat. Mech. Anal., 1 (1958) 410

[166] S. Onogi, T. Masuda and K. Kitagawa,
Macromolecules, 3 (1970) 109

[167] T. Masuda, K. Kitagawa, T. Inoue and S. Onogi,
Macromolecules, 3 (1970) 116

[168] T.G. Fox, S. Gratch and S. Loshaek in
"Rheology", Vol. 1 edited by F.R. Eirich,
Academic Press, New York, 1956

[169] J.T. Gruver and G. Kraus, J. Polym. Sci., 2
(1964) 747

[170] G. Kraus and J.T. Gruver, J. Polym. Sci., A3
(1965) 105

[171] N.J. Mills and A. Nevin, J. Polym. Sci., A-2, 9
(1971) 267

[172] T. Masuda, M. Takahashi and S. Onogi, Appl.
Polym. Symp., 20 (1973) 49

[173] G.V. Vinogradov and A. Malkin, J. Polym. Sci.,
A-2, 4 (1966) 135

[174] W.W. Graessley and L. Segal, AIChE J., 16 (1970) 261

[175] J.A. Cote and M. Shida, J. Appl. Polym. Sci., A-2, 4 (1966) 135

[176] B.D. Coleman and H. Markovitz, J. Appl. Phys., 35 (1964) 1

[177] J.L. White and A. Kondo, J. Non-Newt. Fluid Mech., 3 (1977) 41

[178] K. Oda, J.L. White and E.S. Clark, Polym. Eng. Sci., 18 (1978) 25

[179] J.R. Van Wazer, J.W. Lyons, K.Y. Kim and R.E. Colwell, "Viscosity and Flow Measurement", Interscience, New York, 1963

[180] P. Junk, J. Wortberg and G. Menges, Personal Communication (1977)

[181] W.W. Graessley, S.D. Glasscock and R.L. Crawley, Trans. Soc. Rheology, 14 (1970) 519

[182] I. Pliskin, Rubber Chem. Technol., 46 (1973) 1218

[183] J.L. White and J.F. Roman, J. Appl. Polym. Sci., 20 (1976) 1005

[184] G.R. Cotten, Rubber Chem. Technol., 52 (1979) 187

[185] J. Meissner, Trans. Soc. Rheology, 16 (1972) 405

[186] R.S. Spencer and G.D. Gilmore, J. Colloid Sci., 6 (1951) 118

[187] J.P. Tordella, Trans. Soc. Rheology, 1 (1957) 203

[188] J.P. Tordella, J. Appl. Polym. Sci., 7 (1963) 215

280

[189] E.B. Bagley and A.M. Birks, J. Appl. Phys., 31 (1960) 556

[190] T.F. Ballenger and J.L. White, J. Appl. Polym. Sci., 15 (1971) 1949

[191] J.L. White and H.B. Dee, Polym. Eng. Sci., 14 (1974) 212

[192] K. Oda, J.L. White and E.S. Clark, Polym. Eng. Sci., 16 (1976) 585

[193] C.D. Han and C.A. Villamizer, Polym. Eng. Sci., 13 (1978) 687

[194] Y. Oyanagi and J.L. White, J. Appl. Polym. Sci., 23 (1979) 1013

[195] J.L. White, R.C. Ufford, K.C. Dharod and R.L. Price, J. Appl. Polym. Sci., 16 (1972) 1313

[196] S. Young, J.L. White, E.S. Clark and Y. Oyanagi, Polym. Eng. Sci., (in press)

[197] E.B. Adams, J.C. Whitehead and D.C. Bogue, AIChE J., 11 (1965) 1026; T.R. Fields and D.C. Bogue, Trans. Soc. Rheology, 12 (1968) 39

[198] C.D. Han, J. Appl. Polym. Sci., 19 (1975) 2413; Rheol. Acta, 14 (1975) 173

CHAPTER 6

PAINTS AND PRINTING INKS

By

J. Mewis

CONTENTS

1. INTRODUCTION

This chapter deals with liquids and pseudo-liquids
which are meant to be applied onto a substrate, and
after application, the coating must change into a
solid film. A large variety of materials can be used
in this manner, including ceramics and metals. In
paints, lacquers or varnishes the film forming material
is essentially organic. The present discussion will
be limited to such organic coatings. The general composition and the production methods of a number of
other industrial products are closely related to those
of paints: printing inks, electronic printing pastes,
artists' colours, putty. Most of the rheological
principles pertaining to paints can be applied to
these materials as well. However, the rheological
techniques might differ, especially for putty. Therefore the latter product has not been considered explicitly in this chapter. Printing inks on the contrary
have been included systematically. Whenever necessary
their specific problems have been treated separately.

There are no clear-cut borderlines between the
materials and problems discussed here and the contents
of some of the other chapters. Some coatings are
applied as a melt and polymer rheometry is applicable
in that case. On the other hand, paints and printing
inks contain dispersed particles. As a result they
have a number of characteristics in common with other

industrial dispersions such as paper coatings, ceramics, processed food products and metal slurries.

A paint is basically a dispersion of pigment particles in a fluid matrix. The latter serves to transfer and to keep together the particles. Such pigment dispersions belong to the oldest man-made materials. Even the early predecessors of the modern industrial products must have been distinctly non-Newtonian [1]. Long before rheology existed as a science, paint manufacturers, amongst many others, had to adapt the mechanical properties of their products. As will be seen later, the flow requirements of coatings are so complex and so stringent that they cannot be met by simply adjusting a Newtonian viscosity. It is not surprising then that paints were among the very first products for which the flow behaviour was tested in detail. The discovery, by Bingham, that they were non-Newtonian is a historical landmark in modern rheology [2].

The rise in the importance of polymers, together with the complexity of dispersion structures, caused rheologists to practically abandon the study of dispersions in favour of polymers. As a result most modern texts on general and theoretical rheology overlook a number of phenomena that are important to large classes of industrial products but that do not occur in polymers. At present there is a growing interest in the application of complex heterogeneous systems. The recent fundamental work that is relevant to the materials at hand has been reviewed in several papers [3-5]. The rheology of paints [6-8] and printing inks [9] has also been summarized.

This chapter starts with an explanation of the general nature of organic coatings. Their composition, their function as well as their general rheological behaviour are briefly outlined. Next, the various reasons for doing rheometrical tests are analyzed. They refer to the production as well as the application of organic coatings, including printing inks. On the basis of the previous steps, the detailed requirements for process control and product development are discussed. Amongst the instruments used, two major groups can be

distinguished: the instruments with flow induced by
gravity and rotatory instruments. The limitations and
the procedures to be used for the various techniques
are considered. Both production control and product
development are covered.

2. THE GENERAL NATURE OF ORGANIC COATINGS

The application of rheometry differs from one class
of materials to the other because the mechanical behaviour and the conditions during processing can be
quite different. Behaviour and the method of processing cannot be changed independently since one determines the other. Both are linked to the composition
which in turn is governed by the final function of the
materials under consideration. Therefore the critical
evaluation of the rheological techniques in coating
technology is preceded by a survey of the functions of
the products involved, of their general composition
and of their mechanical characteristics.

2.1 Function and Composition

By definition, coatings are meant to be applied on a
substrate. The purpose of the operation is to provide
the underlying material with a relatively thin film in
order to change the chemical and/or physical properties
of its surface. Most often the film has a protective
or decorative function. In order to form a barrier
against external agents and also in order to be effective over a sufficient length of time, the film must
normally be continuous and unimpaired. In the case of
printing inks the situation is different. Sometimes
they play a decorative role but the general objective
of printing is to convey information. This entails
the production of a complex pattern rather than that
of a continuous layer as exemplified by the printing
of text and by the image reproduction using small dots
in half tone printing. Clearly a number of cases occur
that are intermediate between film and dot formation.
They do not introduce any new problems and do not
require a separate discussion.

In paint and printing ink formulation, it is not only
the final function of the solid film that must be taken

into account. The composition of the liquid phase
should be such that it can be applied on a substrate
where it is to subsequently solidify. Considering
that the final layer thickness normally amounts to 1 -
100 μm, it becomes obvious that an industrial application process requires close control. The rheology
of the liquid has to be adjusted to ensure the formation of a smooth layer with the right dimensions under
the kinematic conditions of the application method
used. During the subsequent drying stage, additional
flow phenomena take place that have to be controlled
to obtain a film of adequate quality.

It can be seen that the basic problem of coating
formulation is to select the proper components for a
given end use and then to adapt them to allow for a
suitable application. Numerous products can be incorporated to satisfy the various physical and chemical
requirements. Although most of them affect the rheological behaviour in one way or the other, only the
most important categories are enumerated here.

The main components of a normal coating material are:
- the binder, an organic film-forming product. This
  serves to keep all components together in a continuous layer which will adhere to the substrate after
  drying;
- pigments and other solid particles (e.g. fillers,
  extenders). These are incorporated to modify the
  physical or chemical properties of the liquid
  product or of the solid film;
- a solvent is eventually added to dissolve, dilute or
  emulsify the binder to make application as a liquid
  possible.
Besides these main components, a normal formulation
can contain a number of additives. Some of them, like
wetting agents, can have an indirect but considerable
effect on the mechanical behaviour in the liquid state.
Some others, like thickeners or thixotropic agents,
are added on purpose to change the flow. Even uninten-
tional variables like humidity can have serious effects.

When the binder is a macromolecular component, the
coating can become solid-like by evaporation of the
solvent or the emulsion medium. This is called

physical drying. It entails a specific change in rheology during solidification. With the other binders, resins, oligomers or drying oils, a further chemical reaction is involved in the film-forming process.

Normally the mixture of binder and pigments would be solid or at least too viscous for most application methods. Previously it was customary to reduce the viscosity by adding a suitable system of organic solvents. Ecological and safety considerations are changing this practice. Various means are available to reduce the quantity of organic solvents. In some cases the molecular weight of the film forming component can be reduced by adapting the chemical drying process, as in U.V. curing. An alternative solution is to use water as solvent or diluent. Finally higher application temperatures can melt the resin, making the solution stage unnecessary.

In all cases, with the exception of some transparent varnishes and of jet printing inks, all systems contain solid particles, either organic or inorganic. These particles have dimensions around the micrometer range. Hence a number of colloidal phenomena such as flocculation and aggregation occur, resulting eventually in the presence of complex particulate structures. Even without the action of colloidal interparticle forces tight clusters might remain, owing to incomplete separation during the production process. The presence of particles and aggregates is of primary importance in understanding the rheology of paint manufacture and application. It does not only cause a more complex behaviour but also interferes with the measurements.

The presence of volatile components like solvents constitutes another source of difficulties. In most rheometers the sample has a free surface where evaporation can take place. A dried surface layer will render a measurement meaningless. On the other hand one needs to know the flow properties of a coating in the various stages of drying, which is a problem in itself. These points will be taken up later while discussing the measuring techniques.

2.2 General Rheological Behaviour

Paints and printing inks cover a large range of flow
characteristics. As a result, it is impossible to
present a general description in this short section.
Few reviews in this area are available [6,10,11]. In
the literature only fragmentary information is often
given and sometimes the data have been collected on
unsuitable instruments or in a limited range of shear
rates. Occasionally terms like yield value are used
with different meanings. (10's cP)

Industrial dispersions start in the cPa.s viscosity
range. These are systems with low solids contents
used in spraying and dipping. The same categories
together with curtain coatings, gravure and flexo-
graphic inks cover roughly the range between cPa.s and
dPa.s. These limits have only an indicative value,
some particular products, e.g. some dip coatings and
some inks could be different. For the same reason
various authors give different limits [6,9-12].

From the dPa.s level 100's cP upwards, non-Newtonian behaviour
becomes prominent. Shear thinning and thixotropy are
the most important elements. Both phenomena can already be encountered with some of the products mentioned above. They become even more dominant in trade
paints formulated to be applied by roller or brush.
The viscosity at relatively high shear rates ($10^3 s^{-1}$)
is around 0.1 Pa.s [8,13,14]. At low shear rates it
can be several times higher. The apparent viscosity
can increase again several times during the first few
minutes after application either owing to thixotropic
recovery or to evaporation [6,15,16]. Also, letterpress inks for fast rotary presses, like news inks,
have viscosities around 1 Pa.s.

The apparent yield stresses of the trade paints
mentioned above and of similar products can vary between wide limits. For high gloss paints and lacquers
the initial yield stress is kept as low as possible.
For flat paints, latex paints and also for some primers,
higher values are found. The average range is 1-10 N/m^2
but it can reach 100 N/m^2 as well. This latter range
holds also for high build paints which can be applied

in relatively thick layers.

At the upper viscosity limit one finds the "fatty"
printing inks. The more important representatives of
this group are the letterpress and offset printing
inks. Their apparent viscosities at moderate shear
rates (10-100 s^{-1}) are approximately between 10 and
100 Pa.s whereas the corresponding apparent yield
stresses are of $O(10^2 \ N/m^2)$ [17] or higher. Again
some degree of thixotropy is found.

Powder coatings are in a separate category as they
are applied as molten droplets. Temperature and time
are now the important parameters. Under usual application conditions, viscosities are in the range $10-10^3$
Pa.s during the initial curing stage [18,19].

Pigment dispersions also show viscoelastic effects,
which can be attributed to the liquid matrix or to a
particulate structure [5]. Most viscoelastic measurements have been performed on model systems with a
simplified formulation rather than on complex industrial products. Low frequency or static moduli cover
the whole range from near zero to $10^7 \ N/m^2$ [19-22].
The low frequency elasticity is related to the same
structures that determine the yield stress and thixotropy. Hence they depend strongly on the previous
kinematic history [23]. The changing structure also
causes strong nonlinearities [23,24]. The high frequency behaviour is less affected by the complex
structure and shows a relatively simple picture based
upon the viscoelastic behaviour of the continuous
phase [25]. Under steady state shear flow the elasticity appears as normal stress differences, which are
difficult to measure in colloidal dispersions [26-28].
Elongational flow with an elastic contribution occurs
in filamentation phenomena [29] but again measurements
on dispersions are difficult to perform under these
flow conditions [30].

It can be concluded that most organic coatings are
heterogeneous systems that eventually contain volatile
and high molecular components. Their mechanical behaviour includes, to various extents, most known deviations from the Newtonian model: shear thinning,

thixotropy, yield stress and nonlinear viscoelasticity. After applying the coating onto a substrate, important physical or chemical changes occur. As will be seen later, there are still important flow phenomena going on during the drying stage. So we need to know the changing rheological behaviour during this period. From these remarks it should be clear that to make a complete rheological control of a paint or a printing ink is a nearly impossible task. In order to reduce the problem to reasonable dimensions, a critical selection has to be made of the necessary measuring conditions, even for product development. For the purpose of process control, further simplifications will be required.

3. RHEOMETRICAL OBJECTIVES

A first step in establishing a feasible and suitable rheological test programme would be the identification of critical flow situations for the materials at hand. From the different conditions, the relevant rheological parameters and the relevant measuring conditions could then be derived. For paints and printing inks, it is important to discriminate between production conditions and the flow patterns appearing during and after the application. This procedure is based on the fact that the rheological properties of the fluid should be within a given range in order to perform well in a given process. The rheological control then warrants the proper functioning of the tested batch. In practice, rheometry can be employed to solve further problems to those mentioned above. It is known that the mechanical response of dispersions is very sensitive to a large variety of product and process parameters. The sensitivity is explained by the possible differences in microstructure that are caused by all these parameters. The detailed microstructure is not only responsible for the mechanical behaviour but simultaneously affects the other physical and chemical properties. In consequence rheology provides an indirect means to predict various other aspects of product quality.

3.1 Production

In principle, the production of an industrial pigment
dispersion can be divided in three steps:
- premixing;
- dispersion or grinding;
- let down.
During the first step, the premixing, the solid particles are more or less blended in part of the liquid
phase. In this operation, large lumps of particles
are broken down and spread homogeneously throughout
the liquid phase. The result is a liquid mixture or
paste which still has a poor quality but which can be
fed to the real dispersion equipment.

The general principles of mixing are valid here.
This means that the geometry has to be adapted to the
viscosity and to the non-Newtonian behaviour of the
material to be handled. For most liquids with a
relatively low viscosity, sufficiently high shear
rates must be created somewhere in the mixer. If the
propeller is properly designed each volume of liquid
will in time be led through the zone of high shear.
With paste-like products having a high viscosity and/
or pronounced shear thinning, the geometry is more
critical. Special precautions must be taken to ensure
that the whole volume is sheared. This result is obtained by designing the moving elements in such a way
that they reach most of the liquid.

The rheological properties are poorly defined at the
start of the mixing operation. Subsequently they
change continuously. The composition can only be
adapted within narrow limits as the final properties
are imposed by the following operation. In general,
shear thickening is avoided because it is difficult to
control and can be dangerous for the equipment.
Excessive shear thinning usually reduces the mixing
efficiency. For the premixing itself rheological controls are not common, although they could be useful in
trouble-shooting. The properties of the final product
are more important and they determine the performance
in the second step of the production.

After an adequate mixing of the components has been

realized, the remaining aggregates must be separated
thoroughly so the liquid vehicle can wet the surface
of the individual solid particles. This step of fine-
ly dispersing the solids and enveloping them with
vehicle is the most essential part of coating production. An insufficient dispersion step may well be
detrimental for all product properties. The operation
is based on a combination of crushing and shearing
actions on the aggregates. To that end various devices
can be used depending on the flow properties of the
premix. The basic principle is to squeeze the aggregates between moving elements like cylinders or
spheres. The motion is imposed by rotation of the
elements themselves, by gravity or centrifugal forces.

The efficiency of a dispersion process depends
strongly on the flow properties of the liquid [31]
although other elements must be considered also.
Patton [8] has presented a general discussion of
several processes whereas Parfitt has treated the
phenomena involved [32]. Only the rheological consequences are considered here. Too low a viscosity will
normally reduce the dispersion efficiency and will
cause excessive wear of the moving bodies. If the
viscosity becomes too high the shear rates might
decrease drastically, especially if gravity or centrifugal forces are involved. The result will be a
reduction in efficiency. The changes of the properties
with time should be taken into account in efficiency
studies. This includes a viscosity that first increases then decreases with ongoing dispersion [6,9].
For most dispersion techniques, rules are available to
estimate the optimal viscosity and optimal operating
conditions.

Fig. 1 illustrates the flow pattern in three typical
processes. In a roll mill the liquid is squeezed
between rollers, rotating at different speeds. Normally there are three rollers. The active sites are
the nip areas which generate locally high shear rates
estimated to be $10^5 s^{-1}$ and higher. Except for the
dispersion efficiency, the transfer from the slower to
the faster roller is important. This factor is also
determined by vehicle viscosity and liquid properties
[9]. However, the flow is complex and difficult to

292

simulate on a rheometer. The same holds for the other
dispersion devices.

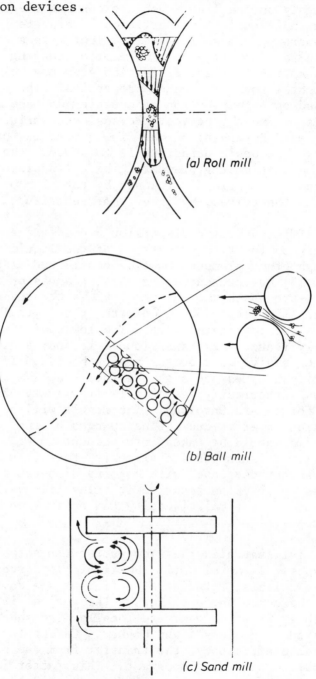

(a) Roll mill

(b) Ball mill

(c) Sand mill

Fig.1 Flow patterns of typical dispersion equipment.

A ball mill consists of a rotating drum with a horizontal axis. Approximately half the volume is filled with liquid and grinding medium. The latter consists generally of metallic or non-metallic spheres. While the mill rotates the spheres make a cascading motion. This motion causes shear flow in the liquid between spheres and crushes the aggregates. Again the flow is complex, shear rates are lower and the dispersion takes much longer than in a roll mill. However, there are several practical advantages that make ball mills attractive in various cases.

Contrary to the two previous designs, sand mills can function in a continuous operation. The grinding medium has smaller dimensions than in ball mills. A central shaft with disk impellers rotates at high speeds, $O(10^3$ RPM). The mixture of beads and liquids is sheared by the rotation but is also thrust against the wall owing to centrifugal forces. Again the liquid is sheared between the beads during their complex motion. Here not only the rheology of the basic dispersion but also that of the dispersion/sand mixture is important in guaranteeing high shear and efficient dispersion.

Premixing and dispersion can be combined as in roll mills or in high-speed mixers. The latter can be used with easily dispersible pigments. A simple impeller at high rotational speed is then sufficient to produce a good dispersion if the rheology of the fluid permits efficient mixing.

The let down step adapts the optimal dispersion composition to the optimal application composition if these two differ. An inadequate let down will lead to a final product with poor quality and with modified rheological properties. The physiochemical phenomena that govern this process are outside the scope of the present treatment [8,33].

As a conclusion it can be stated that coatings are submitted intermittently to high shear rates during the production process. Efficiency requires a mutual adaptation between the rheological properties of the coating and the type and conditions of the process.

The adaptation includes the viscosity of the vehicle, the average viscosity level of the dispersion and its non-Newtonian behaviour. In general shear thickening is avoided but pronounced shear thinning is also a disadvantage. The industrial approach is mainly empirical. Usually, only very elementary flow measurements are used.

3.2 Application

A coating is formulated with a particular end use and a particular kind of application technique in mind. Both product development and production control include rheometrical techniques to ascertain correct flow during the application stage. These methods are selected on the basis of their correspondence with the dynamic conditions during spraying, dipping, brushing, roller or curtain coating or the various printing processes.

The detailed mechanics of coating operations have not been completely analyzed as yet, at least for realistic rheological models. However, the shear rate ranges can be estimated which is important for selecting relevant test conditions [34]. The highest shear rates are probably encountered in the nozzle of spraying equipment with estimated values of over $10^6 s^{-1}$ [35,36]. In roller coating and printing, a similar order of magnitude can be found with hard rollers. With rubber coated rollers, the maximum value turns out to be a few orders of magnitude lower [37], which is still large for measuring purposes. There has been some controversy about the shear rates in brushing [11]. At present it is generally accepted that values of $10^3 s^{-1}$ can be reached [8,13,14]. Processes like dipping involve much smaller shear rates.

A rheological control relevant to the application conditions is required to make sure that the corresponding flow conditions are satisfied. These could refer to good atomization in spraying, brushability and non-dripping in brushing, avoiding roller fly-off and sufficient layer thickness in dipping. Some of the industrial flow conditions are non-viscometric and their simulation in a laboratory device which gives

fundamental characteristics constitutes a non-trivial
problem.

Once the coating has left the application device,
the flow does not stop immediately. If the substrate
is not in the horizontal position the applied layer
will be able to flow under gravity as long as it remains liquid-like. If the gravity flow proceeds too
far, failures such as sagging and curtaining occur.
An instantaneous solidification would not be acceptable either, for each application mode entails its
own surface pattern, which should smooth out before
the drying is completed. Surface tension provides
the driving force for this levelling out. Both
gravity and surface tension cause shear stresses and
shear rates which are orders of magnitude lower than
the maximum values during application. From the
weight of the film, the stresses responsible for sagging can be estimated to be of the order of 1 N/m^2.
From an analysis of levelling for Newtonian fluids
the shear stresses can be estimated [38]. Maximum
values of the order 1-10 N/m^2 have been suggested [16].

The flow conditions after application are clearly *spreading*
different from the earlier ones. The shear rates are
smaller and, from the stresses, values up to $O(10s^{-1})$
[16] have been calculated; all lower values will
occur also. Together with the results of the previous
section, this means that a range of several decades in
shear rate must be controlled. For time-dependent
materials the equilibrium rheogram becomes insufficient
and the shear history must be taken into account. In
addition, if the coating changes during the levelling
period, i.e. 1-2 minutes after application [16], this
aspect must be incorporated in the control. Simultaneously some low shear flow phenomena, linked to
deaeration [39,40] and Bénard cell formation [8,41]
can occur in the film together with segregation of
particle mixtures [8]. Surface rheology is an important factor in some of these aspects of film formation
but has not yet been thoroughly investigated in this
industry [12,34].

The application of printing inks brings along
different requirements from that of paints. Generally

296

speaking a distinction can be made between a distribution phase and a transfer phase. The distribution must provide the printing plate with a continuous layer of printing ink at a constant thickness. Sections with low and high shear rates can be distinguished. The ink duct or press fountain constitutes the ink reservoir. The motion of the fountain roller controls the feeding rate of the press. To ensure a correct feed the liquid must follow the roller. This is a process governed by low shear rates. If the low shear viscosity under the prevailing kinematic history is too high and if the liquid is too shear thinning, it sits back in the ink duct and no material is carried along by the roller [9,42].

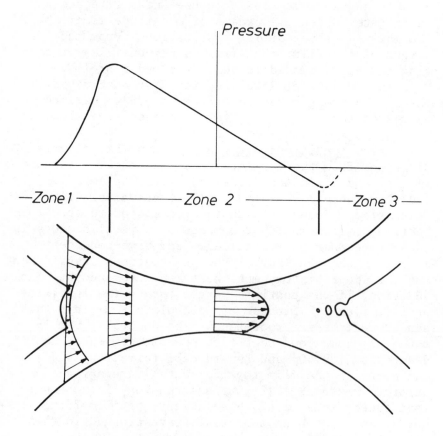

Fig.2 Kinematics of nip flow.

Once the printing ink is in the distribution system, usually a battery of rollers, only nip flow is important. The kinematics of roller passage are illustrated in Fig. 2. In a first zone, pressure builds up while liquid is pushed back out of the nip. Then a squeezing flow zone develops which is associated with a pressure drop. In the diverging part of the squeezing flow the relative pressure becomes negative causing a film splitting zone with cavitation and formation of filaments that finally rupture. The highest shear rates are calculated to be of $O(10^3 - 10^4 s^{-1})$ [37]. The liquid properties must make it possible to smooth out film irregularities as fast as possible.

A general correlation with rheological properties is still lacking. Another flow problem arises when the ink filaments in the exit zone produce droplets upon splitting because this will fill the surroundings of the press with ink droplets ("misting"). The misting tendence of a vehicle can be approximately estimated from its viscoelasticity. [43].

The transfer of the ink from the plate to the paper or to another substrate constitutes a complex flow pattern with corresponding rheological requirements. The tensile forces in the liquid during film splitting should not be too high or the ink transfer will be reduced or the paper might even rupture. In any case, the contours of the printed areas should remain well defined and the dot size should be correct for reasons of image quality. This corresponds to a sufficiently high viscosity, especially at low shear rates. Also the filaments should not be large or they would show up in the printed image. Further, small dots should be printed evenly without the ink being pressed out of the central part. The optimal viscosity depends on the printing process under consideration. In plate cylinders for gravure printing, the ink is contained in a negative relief whereas in letterpress printing, the printed areas are raised above the plate surface. In offset printing there is no relief at all. In screen printing the ink is squeezed through a screen. The forces on the ink layer and the transfer mechanisms are different in the various systems and further depend on the printing speed and on the nature of the substrate.

After transfer, the rheological properties remain important. The printed image must dry as fast as possible to allow for further handling. In wet-in-wet printing, the ink layer should be viscous enough to ensure a suitable transfer for the next ink. If the substrate is porous, penetration during and after transfer must be controlled. Insufficient or excessive penetration will worsen the quality. Surface defects, mottle or speckle, are encountered occasionally which are similar to those encountered in paint films if levelling is inadequate.

It can be seen that the rheological requirements of printing are numerous. As in paints they refer to divergent levels of shear rate: from a few to thousands of reciprocal seconds. The situation for printing inks is complicated by the absence of any stationary shear flow. The deformations are often of short duration. In addition, the flow is non-viscometric. It is not surprising then that the classical laboratory measurements do not correlate well with the various aspects of printing press behaviour.

3.3 Interaction Studies

Up to now only rheological requirements have been considered. Rheological measurements can be extremely useful in other areas of coating technology because of their sensitivity to the microstructure of the dispersion. Some changes in the raw materials, in their conditioning or deviations in the dispersion or let down stages will affect the product quality and will show up in the non-Newtonian behaviour as well. Rheological measurements are especially susceptible to changes in wetting, dispersion quality and colloidal stability. Therefore they can give a fast feedback in product development or an early warning in production control. In addition, the properties mentioned will depend on the detailed production procedure. Hence the latter can be investigated in the same manner [7].

The number of possible applications of rheology in this area is almost unlimited. It will depend on the available insight in the system at hand and on the knowledge about the relation between structure and

properties. The wetting and ease-of-dispersion of
binders or pigments can be derived from a measurement
on a sample prepared according to a standard procedure.
The values of high and low shear viscosities will reflect eventual changes in vehicle-pigment interaction.
Thixotropy could suggest the presence of weak and reversible flocculation. The development of a dispersion
process could be followed by means of rheological parameters [6]. In the early stages of mixing, both the
high shear viscosity and the apparent yield stress will
increase eventually to decrease at a high quality or
degree of dispersion. In the same manner, the influence of various additives or components on vehicle-
pigment interaction could be estimated by means of
comparative measurements [6,7,44].

In a few situations, more quantitative information
can be derived from the data. At a well defined pigment concentration, a number of properties of the
paint, both in the liquid and in the solid state,
change rather suddenly [8]. This critical pigment
volume concentration (CPVC) reflects the closest packing of the pigment particles and constitutes an important characteristic of the system. With a viscometric
technique, it is found as the concentration at which
the viscosity goes to infinity.

As a second quantitative result, the thickness of
the absorbed binder layer on the pigment can be derived,
at least in principle. If a pigment has been dispersed
completely, the high shear viscosity still changes
depending on the quantity of absorbed resin. The
latter will give an apparent increase in pigment volume.
For spherical pigments the thickness of the absorbed
layer could be calculated from the steady state or
dynamic viscosity [45,46]. The fact that the technique
is only applicable to fully dispersed and spherical
particles is a serious limitation.

The properties discussed in this section are all related to particle-vehicle and particle-particle interactions. They will affect film properties such as
colour, gloss, mechanical strength and porosity. Supplemented with other test results, viscosity data can
help the paint formulator predict changes in the

behaviour of the derived film.

4. SINGLE-POINT/GRAVITY FLOW METHODS

4.1 Range of Application

In the paint and printing ink industries a large variety of products are often manufactured in the same plant, each encompassing several production steps. Therefore there is a need for inexpensive, robust and easy to use control devices. The same holds for the industrial users, who normally have neither large laboratories nor qualified personnel. This situation explains the widespread use of single-point instruments. A more complete control procedure would require a more expensive and more delicate instrument, a prolonged control and also a more sophisticated interpretation. As a result, single-point methods are continuously being used on non-Newtonian fluids. If certain conditions are satisfied this procedure can still be acceptable. For Newtonian materials no problems exist. For slightly shear thinning fluids the possible errors are often negligible but even stronger deviations can sometimes be handled.

Two arguments can be used to defend the extended use of single-point methods. One is valid for all instruments, the other only for a particular class of instruments. In production control only similar materials are compared with each other. Therefore it could be argued that, provided an adequate control of the raw materials is executed, the non-Newtonian nature would not change too much. If the viscosity is adjusted at an average shear rate it might then be reasonably close to the preset values at other shear rates. In practice this reasoning is followed often implicitly because viscosity corrections are often made with a single component. In this manner the shear thinning cannot be adjusted independently from the viscosity level. The second argument in favour of single-point measurements refers to instruments with a well-defined shear rate range. For applications where a particular shear rate range is the most critical point, one could limit the control to this range. The selection of test equipment might then be based on the frequency of difficulties or complaints. The limitations of such a procedure should be realized.

In the present section single-point devices are discussed where the material flow is induced by gravity. This group includes the consistency cups and a number of other methods with rather complex kinematics.

4.2 Consistency Cups

Gravity flow through an orifice constitutes one of the oldest and most widespread techniques used in industry to estimate consistency of paints. More than forty different types have been proposed [8,47,48]. The main design variables are the shape of the cup bottom and that of the orifice. The earlier models usually had a flat bottom. This simple shape gives rise to pronounced entrance effects in the orifice. As the latter is usually very short, a zone of fully developed laminar flow hardly exists at all. This situation causes non-linearities in the time-viscosity relation and reduces the sensitivity at low viscosities. With the exception of Oesterle's design [7], the SVLFC cup standardized in Switzerland, the flat bottom design has been abandoned. This instrument has a profiled orifice entrance to smooth the flow.

At present the conical bottom is most frequently used and is encountered in paint standards in numerous countries. The well-known Ford and ISO cups illustrate this category (Fig. 3). A few instruments like the Zahn and Shell cups have a spherical bottom. This type is mainly used for viscosity measurements directly out of the paint container.

With all cups, the efflux time measures the viscosity. Normally the total time required to empty the cup is used but eventually a smaller volume is taken as reference. The efflux time is determined by the orifice geometry. Invariably the orifice has a cylindrical cross section. The length over diameter ratio, the aspect ratio, can vary considerably. The effect of the dimensions can be estimated from capillary flow theory [49]. Sometimes the cylinder is preceded by a conical section, as in the Ford cups, to reduce entrance effects.

302

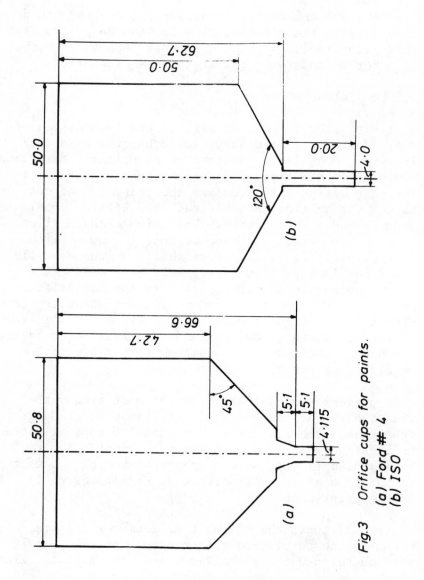

Fig.3   Orifice cups for paints.
(a) Ford # 4
(b) ISO

A high aspect ratio improves the smoothness of the flow and the linearity of the time-viscosity relation. This extends the range of the instruments to lower viscosities. At the same time cleaning might be somewhat more difficult. The average viscosity range of the instrument is adjusted by changing the absolute dimensions of the orifice. In practice the diameter usually amounts to a few millimetres and the aspect ratio changes roughly between 1 and 4. As a result, kinematic viscosities (v) between 2×10^{-5} and 10×10^{-5} m^2/s can be measured at efflux times between 20 s and 2 min. As an illustration, the kinematic viscosity relations for the Ford cup according to ASTM D1200-70, and for the ISO cup, according to ISO 2431-1972, are given:

Ford cup No.2 (range: $2.5-12 \times 10^{-5}$m^2/s):

$$v = 2.388 \times 10^{-6}t - 7 \times 10^{-3}t^2 - 5.7008 \times 10^{-5} \qquad (1)$$

Ford cup No.3 (range: $4-22 \times 10^{-5}$m^2/s):

$$v = 2.314 \times 10^{-6}t - 1.52 \times 10^{-5} \qquad (2)$$

Ford cup No.4 (range $7-37 \times 10^{-5}$m^2/s):

$$v = 3.846 \times 10^{-6}t - 1.73 \times 10^{-5} \qquad (3)$$

ISO cup (range 2.16×10^{-5}m^2/s):

$$v = 1.37 \times 10^{-6}t - 2 \times 10^{-4} \quad . \qquad (4)$$

Equations (1) - (4) are of limited use in the paint industry. Both the instrument and the paints are responsible for deviations. The efflux time in capillaries changes with the fourth power of the radius. With a radius of ± 4 mm, as in the Ford No.4 and ISO cups, the slightest deviation in machining becomes noticeable. Therefore the individual cups have to be calibrated for accurate viscosity measurement. In addition the change in diameter owing to wear cannot be neglected and requires a repeated calibration. For the same reason incomplete cleaning will be a source of errors.

The second restriction on the conversions of eqns. (1) - (4) lies in the non-Newtonian behaviour of many paints. The shear rates in the orifice change from

zero in the centre to a maximum value at the wall.
Further, they decrease with time while the cup empties
itself. As a result the efflux time cannot be converted into absolute units of kinematic viscosity nor can
it be compared with the results of other types of instruments. If an attempt is made to find a conversion,
different results are obtained for different kinds of
products [50].

Still, a cup could be used to verify the reduction in
consistency to a preset level in dilution operations.
With some materials, difficulties are encountered in
establishing the final point where the jet breaks up.

Thixotropic samples are especially susceptible to
errors. Here, the kinematic history will affect the
results. In principle a time-volume curve could provide some information about the combined result of
shear thinning and thixotropy [51]. A more practical
technique would be to compare efflux times immediately
after shearing the sample and after a period of rest.
If the difference between the two results is too large,
a paint cup should not be considered a meaningful control device. Short-time recovery cannot be detected
in this manner. In any case thixotropy reduces the
reproducibility of the cup method [51]. In most standards the kinematic history is prescribed in order to
minimize errors related to thixotropy. Whenever this
property is relevant, more suitable instruments should
be used if false conclusions are not to be drawn from
the data.

Finally, it should be mentioned that consistency cups
are not equipped with a thermal control unit. Nevertheless a correct temperature setting is required in
any viscosity measurement. Tolerances vary depending
on the required precision. Deviations should be
smaller than $1^{o}C$ and, for a precision of a few per cent,
should be $\pm 0.1^{o}C$. If the sample and the cup have been
brought beforehand to the required temperature, their
thermal capacity allows a measurement in ambient conditions deviating up to $5^{o}C$ from the set value. The
effect of temperature on viscosity is often overlooked
in the coating industry. Lacking a suitable regulation
one can always use a temperature correction. If the

cup is used to estimate the amount of solvent that has
to be added, the temperature difference can be incorporated into the viscosity-solvent volume relation [48].

4.3 Other Gravity Flow Instruments

Besides the consistency cups, the paint and printing
ink industries use a number of other single-point instruments based on gravity flow. Some of them have
essentially been designed for Newtonian fluids, others
give a general picture of the non-Newtonian behaviour
under low shear rates. Among the alternative methods
for nearly Newtonian liquids, the falling sphere and
the rising bubble methods must be mentioned. In both,
the net velocity of a foreign object submitted to buoyancy and viscous friction forces is measured. Contrary
to the paint cup arrangements, the present methods
measure the dynamic (η) instead of the kinematic (ν)
viscosity. Although their use can be extended to pig-
mented systems, they are mainly used for transparent
fluids.

The falling sphere method is based on Stokes' law.
The time t is measured which a sphere of radius r
requires to drop a distance ℓ under gravity forces in
a liquid. If the density difference ($\rho_s - \rho_1$) between
sphere and liquid is known, the viscosity is given by

$$\eta = \frac{g}{4.5} (\rho_s - \rho_1) r^2 \frac{\ell}{t} \, , \tag{5}$$

where g is the acceleration due to gravity.

Equation (5) is based on the assumption that the container is large enough to avoid wall effects. In
industrial practice one prefers to work with a small
sample. The presence of the wall is taken care of by
introducing a correction factor which changes with the
ratio of sphere radius r over tube radius R :

$$\eta = \frac{g}{4.5} (\rho_s - \rho_1) r^2 \frac{1}{t} [1 - 2.104 \frac{r}{R} + 2.09 \left(\frac{r}{R}\right)^3 \dots] . \tag{6}$$

By using spheres of different radii and/or different
densities different viscosity ranges can be covered.

As with a cup, the shear rates change throughout the
sample making a comparison with other instruments
impossible for non-Newtonian fluids. In principle
data could be obtained which reflect different shear
rate ranges if several spheres are used. In reality
the conversion to viscosity for non-Newtonian liquids
is too complicated to be of any use. Again thixotropy
might show up in repeated measurements or after changes
in kinematic history.

In the rising bubble method, buoyancy works in the
same direction as it would in spheres with a density
lower than that of the liquid. If an air bubble is
used the flow pattern is different and consequently
eqns. (5) and (6) cannot be used. In practice the
bubble will not even be spherical because it contains
too much air to fit as a sphere in the tube. Lacking
a theoretical formula, the conversion from bubble rise
time to viscosity is based on calibration with fluids
of known viscosity. If the tube diameter, the length
ℓ and the bubble volume are selected properly, the rise
time t can be made to have the same numerical value as
the viscosity, at least at a specified reference temperature [52]. In the paint industry, the viscosity
is sometimes estimated from a comparison of rising time
against that of a set of standard materials [8,52,53].
Sometimes the Gardner-Holdt standards are used [52,53].

With both the falling sphere and rising bubble instruments temperature control is essential. Both have
complex shear rate distributions and should only be
used for quasi-Newtonian liquids. The viscosity range
that can be measured with them is nearly unlimited.

The methods mentioned up to now do not attempt to
simulate relevant kinematic conditions during production or application. Even with the more sophisticated
instruments the situation after application is difficult to imitate. Some single-point instruments have
been devised to provide some information about this
particular aspect. There is a whole range of flow-out
tests [53]. They consist of letting a given amount of
liquid flow on horizontal or inclined planes and measuring the distance the liquid front has moved before
solidification. If a small amount of sample is applied

on a vertical surface sagging is simulated [8,54]. It measures the global effect of the viscosity changes at low shear stresses after application. Such a test clearly demonstrates the effect of thixotropy on sagging. It has an intermediate position between rheological and real application tests.

Over the years, numerous attempts have been made to measure the viscosity changes in the drying film. The effect of drying has been followed by scraping the coated layer from its substrate after increasing lengths of time and measuring the samples in a viscometer [55]. In situ measurements have been obtained by sliding or rotating objects over the film surface and tracing the speed as a function of time. A full analysis of a sphere rotating down an inclined coated plane is not available. Steel spheres of ± 3 mm have been used. For latex paints an angle of inclination of 15-30° has been suggested [16]. For stoving enamels angles of 60-75° have been preferred [39]. In both cases a good correlation between the viscosity-time curve and levelling could be obtained.

The shear rates in the rolling sphere method are variable and difficult to estimate. Quach and Hansen [16] suggest values of about 50 s^{-1} but this estimate is based on sliding instead of rolling. In any case the conditions can be adapted to produce shear rates similar to those of levelling. A good correlation therefore is not surprising. However, the kinematic history is still very complex and calculation of fundamental rheological characteristics is impossible. As an inexpensive and relative method the test has its merits in product development. Recently the sliding plate method for viscosity measurements has been revived [56]. Here the shear rate is better defined but the kinematic history is still complex for thixotropic materials.

Both methods operate at given stress levels rather than at given shear rate levels. This aspect might be an asset as the stresses change less than the shear rates during levelling [34].

5. ROTARY INSTRUMENTS, STEADY-STATE MEASUREMENTS

The instruments of section 4 are unsuited to solve a number of important problems. A considerable amount of coatings are formulated to be non-Newtonian. Consistency cups or similar devices do not provide any relevant information about processing or application conditions of such materials. Instruments are needed with well determined shear rates which are variable over a large range. Rotational instruments with a variable speed drive seem to be a logical choice. Various commercial devices either have been designed for use on pigment dispersions or at least can be used on these materials. They have been tabulated by Mewis [57]. However, a number of specific difficulties arise which require careful consideration during selection and actual usage.

5.1 Couette and Cone-and-plate Geometries

The coaxial cylinders or Couette geometry is suitable for many pigment dispersions. The presence of a disperse phase requires the gap between the cylinder to be sufficiently larger than the particle and aggregate dimensions. A factor of 10 is usually applied in this respect. The restriction is not critical for paints or printing inks except for systems which develop a three-dimensional network after cessation of flow. When shearing starts again the network has to break down and large lumps might be present temporarily. With disperse systems there is always a risk of concentration gradients throughout the gap. At high concentrations and plastic-like behaviour, slip near the wall might occur, whereas at lower concentrations the particles might still migrate to certain zones [58,59]. Sometimes corrugated cylinders are used to avoid slip at the wall. Pigment migration could interfere with transient measurements as it gives rise to an apparent thixotropic effect. Sedimentation is a special kind of migration which must be taken into account when planning measurements. If severe sedimentation should be noticed no rest periods or even no measurements at low shear rates should be made. Special devices have been suggested for the measurement of sedimenting systems but they are not normally used in the coating industry. Because

of the presence of volatile components, evaporation
might occur at the surface of the sample. The total
length of time a sample is kept in the instrument
should be adapted to its volatility. Normally a qualitative verification of the condition of the surface
layer will give a sufficient indication. The limitation of the test duration can be serious in the
characterization of thixotropic liquids. If the space
above the surface of the sample can be isolated, it
can be filled with air saturated with the volatile
component. Few commercial viscometers are designed
for this purpose.

Besides the composition of paints and printing inks,
their rheological behaviour warrants some caution.
Strongly non-Newtonian systems especially, with a yield
value or pronounced thixotropy, prove to be difficult
to deal with. In a Couette viscometer, the shear
stress changes inversely proportionally to the radius.
If the sample is strongly shear thinning the shear rate
distribution can become very broad in some commercial
instruments. As a result the simple equation to calculate the apparent viscosity cannot be used. One has
to take into account the shear rate distribution and
consequently use the more complicated formulae [60].
The latter procedure is not necessary for relative
measurements. Still, the instrument remains unsuitable
for transient measurements on thixotropic systems as
each layer has a different shear history. If a general
expression for the time dependency were available one
could try to incorporate it in the calculations. However, no such expression is available to date [61].
Theoretically, a yield value can be taken into account.
With real coating materials, the yield stress as well
as the rheogram are time-dependent and the theoretical
procedure does not work. One then has a slip layer of
unknown thickness from which no fundamental characteristics can be calculated.

Finally, the Couette geometry has practical drawbacks
for some coatings. Filling might be difficult with
very viscous or plastic substances and preventing
entrapped air in the sample is a problem. Despite the
limitations and drawbacks, coaxial cylinder viscometers

are being used successfully in the analysis of many
pigment dispersions.

The equilibrium viscosity curve provides information
on the shear thinning behaviour. If the instruments
cover a sufficiently large range of shear rates, the
viscosity during and after application can be obtained.
If not, an extrapolation of the data to somewhat higher
shear rates is normally possible without too many risks,
especially if viscosity curves are known for dispersions with lower concentrations of pigment or thickener.
These curves do not normally cross each other; if the
extrapolated curves do, they should be disregarded.
Viscous heating is a possible source of error [62].
For some categories of paints, a master curve technique
with horizontal and vertical shift factors has proved
to be helpful in extrapolation [62]. Extrapolation to
lower shear rates is more risky than to higher ones
because a Newtonian plateau might develop or on the
contrary the sample might be plastic. In addition, the
equilibrium viscosities at low shear rates do not correlate so well with levelling behaviour [34]. Some
special measuring procedures have been suggested. They
are applicable on other rotary rheometers as well and
will be discussed after cone-and-plate instruments have
been considered.

The cone-and-plate geometry has several advantages
over the Couette geometry. The small sample size, the
homogeneous shear rate and the possibility of an immediate conversion of the data to apparent viscosities are
attractive features (cf. Chapter 1). However some
specific problems arise with industrial pigment dispersions. As with coaxial cylinders, the ratio between
gap and particle dimensions must be respected. Here
also concentration gradients can interfere with the
measurements [64] as well as evaporation at the free
surface of the sample. The instrument is in principle
suitable for relatively high shear rates because of the
small cone angles. The commercial instruments used on
paints and printing inks have cone angles between 1/3
and 4°. Care must be taken that, at high shear rates,
no viscous heating takes place [65]. Especially with
printing inks this might impose serious limitations on

the maximum shear rate. It is also true that the
thermal control on commercial instruments is not always
optimal. Even before a limit is reached for thermal
reasons, pigment dispersions often give rise to shear
fracture. This phenomenon consists of a constriction
at the free surface induced by the flow (Fig. 4). With
increasing shear rate the effect becomes more pronounced. Material creeps out of the gap onto the rotating
member while the air gap between cone and plate pro-
gresses towards the centre. As a result, the torque is
transmitted through a cross section of liquid that is
much smaller than the cone surface area. Material
functions calculated from such data must be in error.

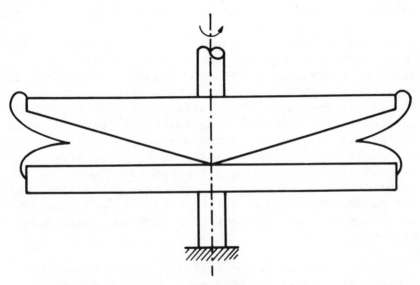

Fig.4 Shear fracture.

While performing measurements on dispersions it is
essential to know whether shear fracture occurs and if
so from what shear rate on. Unfortunately no theoretical technique is available to predict the appearance
of shear fracture. Hutton [66] found a correlation
between the elasticity and the critical shear rate of
an homologous series of polymer fluids. When various
dispersions were included, tests in this laboratory
[67] did not show any correlation with viscoelastic

properties. If only the pigmentation level is changed, the critical shear stress seems to remain relatively constant as can be seen from the upper values in the data of Mewis and de Bleyser [26]. However, the published results [66-69] do not lead to a general reduction scheme.

Lacking any prediction method one has to rely on visual inspection to determine the critical conditions for shear fracture. If the cone and plate are surrounded by a thermal chamber the control can disturb the thermal equilibrium as the chamber has to be opened each time. No effective ways to eliminate shear fracture are known. Some degree of reduction is obtained by improved alignment and by using an excess of liquid.

With some commercial cone and plate instruments the relative position of the two elements is secured by means of a spring. With some viscous dispersions, the gap still opens against the spring under shear, ruining the sample geometry. The latter will also depend on the quality of the machining during the production of the plates and the cones and also on positioning. With the small cone angles normally used, minor errors become noticeable as Cheng has shown for commercial instruments [70]. Inexpensive cone and plate viscometers for paint control are available but in each case it should be verified whether they meet the desired precision.

Theoretically, powder coatings can be measured with rotational rheometers in very much the same way as other coatings. In practice, their high application temperature requires an instrument with a high temperature unit. Further, one would like to follow the curing which takes place here in a non-isothermal process. Cone-and-plate rheometers have proved to be suitable [18,19,71]. Viscosity is measured as a function of time at various temperatures from which one tries to predict the viscosity-time relation under the non-isothermal conditions of the real process [18,19]. Normal stress measurements provide additional information about the polymer structure. Oscillatory techniques can be used in a similar way (section 6). Flow through instruments using capillaries [72] or plastographs [73] can be used for comparisons but gives a less detailed picture of the curing process. Printing inks pose the same problems as other pigment dispersions of the same viscosity range. As they can go higher in viscosity the corresponding difficulties of viscous heating, shear fracture, change in gap geometry become worse too. Alternative methods will be dealt with below. However the rotary instruments are difficult to replace for the low shear rates, which are important in the feeding section of the printing press. Some aspects of nip flow have been associated with high shear viscosity and viscoelasticity of the dispersions [9,43] but generally an additional control technique is used to cover nip flow as will be seen later.

Only steady state shear flow has been discussed until now for two geometries of rotatory flow. The problem now arises as to how to use the same instruments to measure the degree of thixotropy. It has been stated in section 2 that no general form of constitutive equation for thixotropic materials is available. As a result one does not have available a well defined set of material parameters that completely characterize thixotropy nor can one convert the behaviour from one condition to another. The further treatment will cover the real time dependency only. The confusing use of the term thixotropy to mean shear thinning [74] still persists [48] but is considered unsuitable. It is true that thixotropic structures will affect the equilibrium viscosities. However no satisfactory general relation between shear thinning and thixotropy has been found. Each thixotropy model predicts an equilibrium viscosity curve but this is different for each model. For this reason no scientific basis is available for calculating thixotropy from the equilibrium rheogram. Defining thixotropy as the deviation from Bingham behaviour, as suggested by Oesterle [7,75,76] is especially open to discussion.

Time dependency implies transient behaviour and this constitutes a suitable route to approach thixotropy. Three different types of transient measurements are used in the coatings industry, two of which are also known in other areas. These and other methods are reviewed in ref. [61].

One of the most common methods consists of increasing the shear rate linearly with time and subsequently decreasing it in the same manner (Fig. 5). The shear stress-shear rate plot of this experiment describes a hysteresis, its surface area has been suggested as a measure of thixotropy [77]. This surface area depends on the maximum shear rate, the acceleration and on the previous history of the sample. Acceleration itself can eventually give rise to hysteresis without any thixotropy.

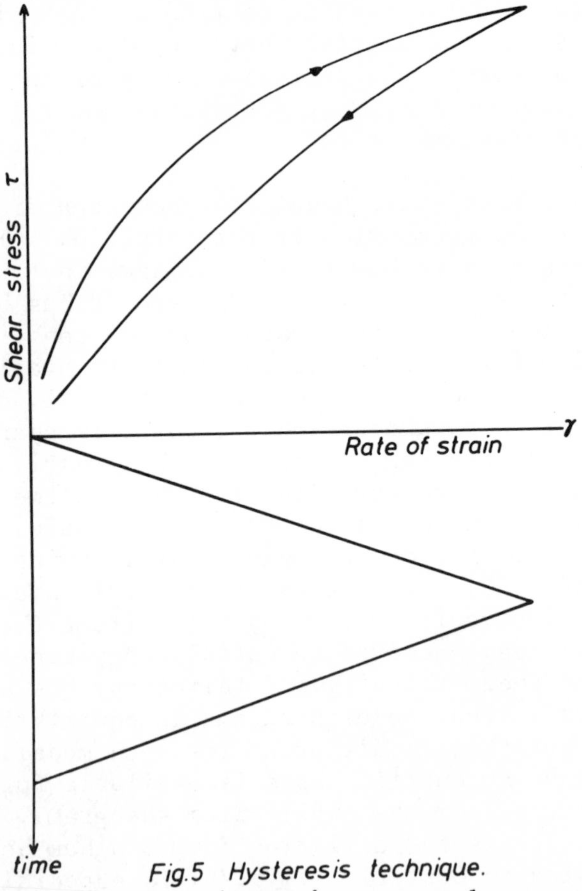

Fig.5 Hysteresis technique.

If the hysteresis experiment is repeated several times, continuously smaller loops are described until an equilibrium loop is reached. The latter does not depend on the shear history any more but is a less sensitive measure than the first loop. Hysteresis loops are difficult to translate into material functions as indicated by Cheng [78].

A practical difficulty is caused by most commercial instruments because they are not equipped with a continuously variable drive. Hence the kinematics of Fig. 5 cannot be reproduced. Instead, a stepwise change must be used following accurately a determined time schedule. The results can be a relative measure of thixotropy at the most. Further, the correlation with the practical behaviour has never been clearly demonstrated [79]. Nevertheless the stepwise procedure has been standardized in certain countries [80]. If enough time is spent at each shear rate the hysteresis surface should become zero when the equilibrium stresses are reached. In practice it can take a considerable amount of time to construct the equilibrium viscosity curve in this manner. Usually, one prefers to follow a procedure whereby shearing at a given shear rate is preceded in succession by periods of higher and lower shear rates (Fig. 6) [81].

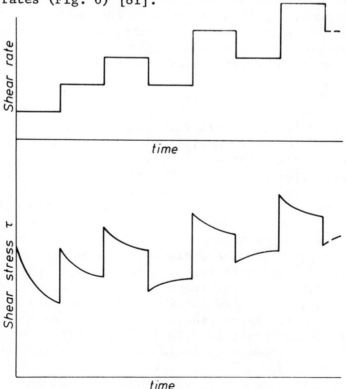

Fig.6 Measuring the equilibrium viscosity of thixotropic samples.

In this manner one hopes to obtain upper and lower bounds for the equilibrium in a short period. In the meantime the reversibility of the time dependent effects can be verified. At low shear rates particularly it takes a long period of time to reach equilibrium. The high shear limit of viscosity and the extent of shear thinning can be used to study the interactions discussed in section 3.3.

Owing to the various theoretical and practical drawbacks of the hysteresis method, alternative techniques have been explored. In principle the stress transients on sudden changes in shear rate should characterize thixotropy. Again it turns out that no general models are available to reduce data from various jumps in shear rate to a small number of model parameters [82,83]. Hence one has to approximate the relevant conditions in thixotropy measurements for product development. This means that in the paint industry one is interested in the recovery under the small stresses of levelling and sagging after shearing at the high shear rates of the application process.

Two techniques have been derived from this principle. The most common one is also applied to other materials. Basically it is a shear rate jump from a finite value to zero in order to measure the recovery of thixotropic structure after the shear flow stops. It approximates the real flow problem but the small stresses during recovery are replaced by zero stress. There is however a difficulty because no viscosity can be detected without shearing. In order to measure the viscosity increase after a given rest period the rotation is started again and the overshoot or peak stress is assumed to represent the recovery value at that time [23,54,84]. The experiment has to be started again from the beginning to obtain a following data point. Alternatively the overshoot can be replaced by oscillatory measurements to detect changes with time if the available rheometer is equipped with this mode of measurement [23,85]. Recovery is a useful and sensitive but time-consuming technique. It measures both the magnitude of the thixotropic contribution to stress and the time scale for recovery. The viscosity increase during the first one or two minutes gives some indication about the behaviour during levelling and sagging. The final recovery measures the storage behaviour [54].

In a second recovery technique an attempt is made to simulate even closer the real flow behaviour. It is a special application of a rather general technique which could be called relaxation viscometry [86,87]. It can be applied on any rotational instrument where the torque is measured as a displacement of the rotor, e.g. through a spring. The rotor is rotated until a given torque or stress is reached. Then it is released and the subsequent displacement is followed as a function of time. Instrumental parameters, a and b, link the total torque T and the angular displacement θ to shear stress τ and strain γ

$$\tau = a T \, , \tag{7}$$

$$\gamma = b \theta \, . \tag{8}$$

The spring constant interconnects torque and displacement:

$$T = c \theta \, . \tag{9}$$

For a Newtonian fluid this leads to the following expression for the relaxation:

$$\theta = \eta \, \frac{b}{a.c} \, \frac{d\theta}{dt} \, . \tag{10}$$

With an initial displacement θ_0 one finds for the movement a linear plot in a semi-logarithmic diagram:

$$\ln \frac{\theta}{\theta_0} = \frac{b}{a.c} \, t\eta \, . \tag{11}$$

For other constitutive equations, other relations can be found analytically or numerically.

The method is meant for inelastic liquids. With thixotropic samples the resulting viscosity-shear rate curve will depend on the kinematic history prior to the experiment and the values of θ_0 and c. Again the data are unsuitable for the calculation of fundamental parameters. They provide only a relative classification. If the liquid has been sheared at high shear rates, the levelling after application is somewhat simulated. Several authors have found a good correlation between such a technique and real levelling behaviour [88-91] if a suitable low starting stress is used. Further applications of this relaxation technique include measuring the thixotropic recovery curve at rest and extending the measuring range of a given instrument to higher viscosities [92].

5.2 Simplified Production Control and On-line Instruments

The instruments discussed in the preceding section can be used and are being used in production control. However, there has been a request for easier to use and cheaper control devices. Some of the instruments made to meet this need retain the geometry of the previous instruments. Eventually they might have only a single-speed drive in which case they replace the single-point instruments with gravity flow of section 4. The present instruments have the advantage that they provide a real viscosity measurement even for non-Newtonian fluids. In the paint industry, special devices are used that operate at a single shear rate of the order of 10^3 s^{-1} to control such application properties as brushability [93-95]. They are designed specifically for the viscosity range of interest in this particular problem. The geometry is of the Couette type. It should be mentioned that high shear rates can also be reached on other instruments, in particular slit [96] and capillary [97] viscometers. These instruments are less convenient to use in production control.

Several types of instruments are being used that are even more simple. They merely consist of a rotating element that is plunged into a container with the fluid to be measured. Some are bench-type instruments, some are portable viscometers. If the rotating member is a cylinder, the results can in principle be

converted into fundamental characteristics [60] . The
conversion presupposes that data at several speeds can
be taken.

   For Newtonian liquids a calibration is always possible. In any case, this group of instruments shows a
very broad shear rate distribution. The connected
problems for shear-thinning and especially for thixotropic materials have already been mentioned.

   Viscometers with rotating elements like bicones,
spheres, disks, paddles and the like can also be used.
For some of them, like spheres [98,99] and disks [100],
conversions to apparent viscosities are available.
The geometries encountered in industry include disks
and paddles. The Stormer-Krebs viscometer especially
has been popular [8,53] (Fig. 7).

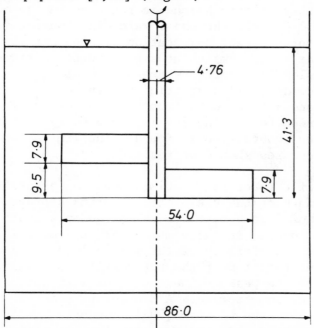

Fig.7 The geometry of the Stormer-Krebs Viscometer.

The rotating element is a two-bladed paddle. Contrary to the other instruments of this category, it is
not driven at constant speed but by a constant torque
which is controlled by a weight. The weight W that
causes the paddle to rotate at a given speed, 200 rpm,
is used as a measure of viscosity. Its value is

derived from a weight-rotational speed diagram. The
experiment can be speeded up by using a stroboscopic
device. The data are often converted to an arbitrary
scale of Krebs Units (KU). A conversion table W-KU
[101] is provided with the instrument. For Newtonian
fluids the following equations can be used to correlate W to KU or to absolute viscosity units. W is
expressed in grams:

$$KU = 17.2(\log W)^2 - 3.9 \log W \quad , \tag{12}$$

$$\eta = 0.0051 \ (W - 34) \quad . \tag{13}$$

In principle several weights could be used to obtain a
qualitative picture of non-Newtonian behaviour in the
same manner as with other rotating elements.

Apparently this technique is not used in the paint
industry. Clearly the shear rate distribution is too
complex for a direct computation of apparent viscosities. The criticisms about single-point instruments
with complex flow patterns are applicable here.
Rowland *et al* [102] have tried to develop approximate
conversions between data from the Stormer and from
other viscometers. Inevitably, for each type of paint
a different conversion is found. The limitations of
such procedures should be realized.

A specific rotor is sometimes used to measure the
gel strength or the yield stress during storage. It
consists of 4 to 6 flat blades fitted on a shaft [103]
If it is lowered in the sample, only the material in
the immediate vicinity of the blades is disturbed. A
subsequent rotation of the rotor will load undisturbed
material. The peak torque can be used as a measure of
yield stress on storage. A Stormer viscometer can be
used for the same purpose [104]. Couette or cone-and-
plate geometries provide more absolute results but
cannot be used for storage tests of long duration [105,
106].

Considering the importance of viscosity measurements
in resin, varnish, paint and printing ink production,
a strong tendency towards automation of this control

would be expected. The variability and complexity of products and processes as well as a number of technical difficulties have delayed automatic viscosity control in these industries. The main applications are in resin reactors and in dilution operations. Viscosity is used as a passive control, for instance to determine the end point of a polymerization process. Alternatively, interaction with the process is also possible: the viscosity measurement can be used to regulate automatically the solvent stream in a dilution step. Among the major difficulties, temperature control and the liquid renewal at the sensor surface should be mentioned.

A process viscometer can be either located in the vessel or reactor itself, in the main liquid stream from the vessel or in a special by-pass [107]. In the first two cases the temperature in the measuring device cannot be deviated easily from the process temperature. This lack of controlled thermal conditions interferes seriously with the measurement, especially in non-isothermal processes. In principle, a temperature reading could be incorporated that automatically corrects the viscosity measurement or regulation. Apparently such a system has not yet been commercialized for the coating industry. If the process viscometer is mounted in the vessel, care must be taken to avoid interferences from the agitator on the measurement. One solution is to measure the torque on the agitator itself [107]. If a separate instrument is used it should be screened from the flow induced by the agitator. A Couette geometry would be effective in this respect but then a sufficient renewal of liquid in the gap should be ensured. The latter is easier to do when the viscometer is situated in a pipe outside the vessel. Both the thermal and renewal problems can be overcome if a by-pass is used to deviate a fraction of the main stream through a thermal bath and a viscometer. The residence time in the thermal bath will slow down the feedback. Delays of about two minutes are given for commercial instruments [108].

With regard to viscometer geometry, various types can be recognized, most of them rotating instruments.

Coaxial cylinders,especially,are frequently used as in
the process viscometers by Contraves, Haake and
Medingen. The rotational instruments are available
with one or more speeds which eventually permits an
automatic control of the non-Newtonian behaviour. In
the next section, falling-rod viscometers will be discussed and a process control device based on this geometry has also been suggested [109]. It does not provide a continuous recording of viscosity but rather a
periodic one.

6. OSCILLATORY AND SPECIAL TEST METHODS

The instruments discussed so far cover most of the
problems encountered with most of the industrial pigment dispersions. Exceptions are the drying stage
after application for all coatings and the application
process of viscous printing inks like the ones used in
letterpress and offset printing. These cases are concerned with high viscosities and eventually with a
transition to solid behaviour. Steady state shear
flow can be used but leads to a number of problems
(section 5). Some of the problems can be overcome if
the materials are characterized by means of oscillatory
flow. In this manner, the industrial flow conditions
are not exactly simulated but comparative information
about highly viscous samples and about the liquid-solid
transition can be collected.

In a number of application processes, the material
is only submitted to shear for very brief periods.
Passage through the nip of rotating rollers in roller
coating or printing illustrates such a transient behaviour. An oscillatory motion might then be as good
an approximation to the real kinematics as steady flow.
Voet seems to have been the first to realize this possibility [110]. It was later demonstrated that
measurements in the kHz range correlate with nip flow
as far as the effect of pigmentation is concerned [25].
The elasticity interferes in a more complicated manner
in the droplet formation and in the effect of roller
speed [43]. However high frequency measurements on
viscous dispersions are subject to some limitations:
the wave length should be larger than the

heterogeneities in the system,and the wetting of the instrument surface should be perfect otherwise the results are completely erroneous. This technique could be very useful in product development and especially in selecting vehicles or resins. Devices for the moderate frequency range [111,112] have been commercialized.

Whenever elasticity is important, oscillatory measurements are preferred for dispersions because "stationary" effects of elasticity such as normal forces are difficult to measure. Care should be taken to discriminate between elasticity associated with the binder and that related to particulate structures. It is the former that is normally the most relevant. Therefore,a separate measurement on the vehicle might be useful in product development.

Oscillatory measurements in the MHz [113] and in the kHz [114] range have been suggested to follow the liquid-solid transition during physical and chemical drying. The high frequency data have only a comparative value. Sometimes it is not possible to calculate even the complex modulus from the data [114]. They have the merit of being applicable during the whole drying process whereby the kHz data seem to be more sensitive in an earlier stage and the MHz data at a later stage. The said techniques have not yet found a systematic use in coating technology. Low frequency behaviour, less than 50 Hz, is very sensitive to structure and is useful in understanding pigment-vehicle interactions. It can also be used effectively to study the curing of powder coatings [19,115]. A peculiar geometry is found in the torsional braid analysis [116]. It consists of a woven glass braid impregnated with the liquid to be measured. The system is used as a torsion pendulum to measure the relative changes in rigidity and damping as a function of time. The technique is suitable to follow drying, especially for chemically drying mechanisms where the layer thickness is not important.

In the printing ink industry there has been a need for a control instrument that is easier to use than the usual rotational viscometers for very viscous materials. It should also overcome the corresponding problems like viscous heating and shear fracture. The falling rod viscometer and similar devices have met this need. As in the Couette geometry the liquid is held in an annular gap (Fig. 8). The motion however is now axial instead of radial. One of the cylindrical walls, usually the inner one, moves under gravity while the other is stationary. The result is a telescopic flow field. In cylindrical coordinates one has for the shear stress and rate of strain

$$P_{rz} = \frac{Wg}{2\pi rh} \tag{14}$$

and

$$e_{rz}^{(1)} = -\frac{dv_z}{dr} \quad , \tag{15}$$

where W is the weight of the falling cylinder.

In a typical design the rod has a radius of approximately 5 mm, whereas the annular gap is of the order of 10^{-2} mm. From eqn. (14) one then expects relative changes in shear stress throughout the gap of 1%. For most practical purposes this stress distribution can be neglected. As a consequence, each data point can be converted immediately to a point on the viscosity-shear rate curve. In this respect one has the same advantage as in a cone-and-plate instrument. The difficulties of shear fracture and viscous heating are eliminated or reduced.

Assuming a constant shear stress in the gap, the velocity profile is derived from eqn. (15) through:

$$dv_z = \frac{P_{rz}}{\eta} dr \quad . \tag{16}$$

Integration of eqn. (16) and substitution from eqn. (14) gives:

$$v_z = \frac{Wg}{2\pi h\eta} \ln\left(\frac{R_u}{r}\right) \quad . \tag{17}$$

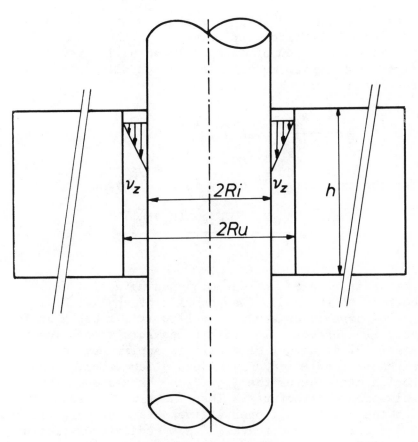

Fig.8 Falling - rod viscometer.

Eqn. (17) can be rearranged to provide upon differen-
tiation a more suitable expression for the shear rate:

$$e_{rz}^{(1)} = \frac{v_z}{\ln\left(\dfrac{R_u}{r}\right)} . \tag{18}$$

Eqns. (14) and (18) can be evaluated at $r = R_i$ to give
a point of the viscosity curve. For the shear rate,
a series expansion of $\ln(R_u/R_i)$ can be limited to the
first term, which means using the parallel plate
approximation:

$$e_{rz}^{(1)} \simeq \frac{v_z(R_i)}{R_u - R_i} . \tag{19}$$

The corresponding viscosity is given by:

$$\eta = \frac{W.g. \ \ln(R_u/R_i)}{2\pi h \ v_z(R_i)} \simeq \frac{W.g. \ (R_u - R_i)}{2\pi h \ v_z(R_i)} . \tag{20}$$

The shear rate is increased by adding additional
weights to the rod. The position of the latter is
not externally ensured. The flow itself has a self-
centering effect. In order to perform a measurement,
a sufficient amount of liquid is applied on the rod
which is then moved up and down a few times. The
motion distributes the liquid and breaks down the
thixotropic structures. Then the time is measured
for the rod to fall over a given distance, e.g. 15 cm.
Shear stresses can reach 10^4 N/m^2 and viscosities up
to 100 Pa.s. can be measured, the upper values of the
shear rate are $\pm 10^3$ s^{-1}. Recommended procedures for
the falling rod viscometers are available [116,117].

The use of the instrument under consideration is
not without its problems. Because of the large free
surface of the sample, volatile components can eventually evaporate during the measurement. Further,
the temperature control is not very effective. At
best, water from a thermostatic bath can be circulated

through the outer cylinder. In this manner only a fraction of the sample is affected directly. Viscous heating is reduced because of the short residence time in the gap but the cleaning and manipulating of the rod can easily change its temperature. Clearly the ambient conditions should also be closely controlled in order to have a reasonable accuracy. The geometry can be another source of errors. The gap width used is typically a few hundredths of a millimetre. Small differences in machining or changes owing to wear could cause relatively large deviations from the design value of shear rate and consequently of the value of the viscosity computed from eqns. (19) and (20). Therefore the use of the design parameters is not recommended; instead one should rely on a calibration with a Newtonian oil. The calibration should be repeated at regular intervals. A precision of ±5% can be reached in production control, which is adequate. Mill has shown that, provided the shear history is similar, the results from a falling rod viscometer correlate with those of a cone-and-plate instrument [118].

The band viscometer is based on the same principles as the instrument discussed above [8,119] but is less frequently used. Instead of a rod, a band now moves through a slit filled with sample. The equations governing flow are those for sliding motion between parallel plates. Again the speed can be changed by adding various weights to the lower end of the band.

Notwithstanding the large variety of rheological instruments, it is felt that the control of printing ink flow is still unsatisfactory. In particular, an easy method to predict the splitting forces or "tack" is still lacking. Originally it was qualitatively appraised by estimating the force necessary to split an ink film between two fingers. A number of mechanical fingers were later introduced (e.g. [120,121]). In these devices the kinematics are complex and do not simulate nip flow. At present rotary tackmeters are extensively used in printing ink technology. The principle of the instrument is illustrated in Fig. 9.

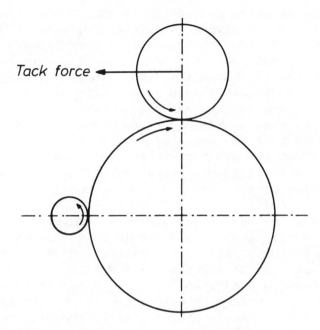

Fig.9 Principle of a rotary tackmeter.

A tackmeter is essentially a system of three rotating rollers. The central roller is driven by a motor, the second roller is positioned on top of the central roller, the third roller describes a reciprocating motion that helps to smooth the liquid layer before it enters the nip between the two other rollers. The liquid on the surface of the rollers transmits forces from one to the other. The external force or torque necessary to keep the top roller in place is used as a measure of tack. The dynamics of nip flow in tack-meters are complex and have not been completely analysed [9,37,122]. The difficulties involved in correlating viscometric data with nip flow behaviour have been mentioned in sections 3 and 5. They explain in part the success of tackmeters in printing ink con-trol, where they are used to supplement the information from viscometers. Tack seems to give better correla-tions with a number of quality aspects than viscometric data, for instance transfer in roll mills, picking of paper, transfer in wet-on-wet printing and press stabi-lity. This case illustrates the fundamental problem in industrial rheometry. How can viscometric data predict non-viscometric flow behaviour? The problem arises mainly during the formulation. In production control the simularity with the real flow can usually be relaxed. An experimental analysis of roller passage has shown some distinct relation with viscoelastic properties [25,43]. This relation deserves further investigation. However, from a practical point of view, a tack measurement might still be the preferred control technique. The main drawback is the influence of instrumental characteristics on the measurement, which limits the conversion from one tackmeter to the other [123].

7. CONCLUSION

A large selection of instruments is available to assist in the formulation and the control of paints and printing inks. Several of the simple control devices neither have well-defined kinematics nor the possibility of varying the shear rates. Hence they have only a limited value in the testing of non-Newtonian materials. Quite generally the behaviour at high shear rates must be known to verify proper behaviour during the application process, for instance at shear rates in the range $10^3 - 10^4 s^{-1}$ or even higher. Rotary rheometers can be used in this particular range but specific problems may arise, which should be anticipated. The same measurements provide some information about the aggregation condition of the solid particles in the dispersion.

The relevant low shear behaviour is more difficult to simulate. First, the shear rates are variable and do not remain constant. Secondly, the thixotropic nature of the materials requires the previous history to be taken into consideration. The hysteresis loop method is used for fast screening or for fast control. The recovery under small stresses or shear rates is a more reliable technique.

Methods are available to measure the viscoelasticity of the dispersions and the changes in viscosity during curing. Both aspects have been shown to be relevant but their application in industry is still very limited.

The same holds for the use of rheological measurements in the prediction of non-rheological properties of the product. Fortunately a gradual improvement can be observed, which slowly penetrates the national standards of several countries.

REFERENCES : CHAPTER 6

[1]   U. König, Chemiker-Z. 92 (1968) 343

[2]   E.C. Bingham and H. Green, Proc. Am. Soc. Test.
      Mat. 19(II) (1919) 640

[3]   B.W. Barry, Adv. Colloid Interface Sci. 5 (1975)
      37

[4]   J.W. Goodwin, Colloid Science Vol. 2,
      D.H. Everett (senior reporter) The Chemical
      Society, London (1975) 246

[5]   J. Mewis and A.J.B. Spaull, Adv. Colloid
      Interface Sci. 6 (1976) 173

[6]   R.N. Weltmann, Rheology Theory and Applications,
      F.R. Eirich (ed.), Academic Press, N.Y.,
      Vol. III, (1960) 189

[7]   K.M. Oesterle and M.B. Palmer, Characterization
      of Coatings, Physical Techniques II,
      R.R. Myers and J.S. Long (eds.), M. Dekker,
      N.Y., 1976, 123

[8]   T.C. Patton, Paint Flow and Pigment Dispersion,
      J. Wiley, N.Y., 1964

[9]   A.C. Zettlemoyer and R.R. Myers, Rheology Theory
      and Applications, F.R. Eirich (ed.), Academic
      Press, N.Y., Vol. III, 1960, 145

[10] E.K. Fischer, Colloidal Dispersions, J. Wiley,
      N.Y., 1950

[11] J.G. Savins, Off. Dig. Fed. Soc. Paint Technol.,
      32 (1961) 1250

[12] G.P. Bierwagen, Progr. Organ. Coatings 3 (1975)
      101

[13] W.K. Asbeck, Off. Dig. Fed. Soc. Paint Technol.,
      32 (1960) 668

332

[14] J.D. Dormon and D.M.D. Stewart, J. Oil Col. Chem. Assoc. 59 (1976) 115

[15] H.K. Fischer, J. Coll. Sci. 5 (1950) 271

[16] A. Quach and C.M. Hansen, J. Paint Techn. 46 (1974) 40

[17] T.B. Turner, Brit. Ink Maker, 18 (1976) 55

[18] S. Gabriel, J. Oil Col. Chem. Assoc., 59 (1976) 52

[19] M.J. Hannon, D. Rhum and K.F. Wissbrun, J. Coating Techn., 48 (1976) 42

[20] M. Takano, Bull. Chem. Soc. Jap. 37 (1964) 78

[21] A.F. Douglas, G.A. Lewis and A.J.P. Spaull, Rheol. Acta 10 (1971) 382

[22] A. Zosel, FATIPEC XIV, Budapest (1978) 731

[23] J. Mewis and R. de Bleyser, J. Colloid Interface Sci. 40 (1972) 360

[24] G. Schoukens, A.J.B. Spaull and J. Mewis, Proc. VIIth Int. Congr. Rheol., C. Klason and J. Kubat (eds.) Chalmers Univ., Gothenburg, 1976, 458

[25] L. Hellinckx and J. Mewis, Rheol. Acta 8 (1969) 519

[26] J. Mewis and R. de Bleyser, Rheol. Acta 14 (1975) 721

[27] J.E. Glass, J. Coating Techn. 50(640) (1978) 61

[28] G. Schoukens and J. Mewis, J. Rheol. 22 (1978) 381

[29] I.G. Thomson and F.R. Young, J. Oil Col. Chem. Assoc. 58 (1975) 389

[30] J.E. Glass, J. Coating Techn. 50(641) (1978) 56

[31] A.H.P. Skelland, Non-Newtonian Flow and Heat
      Transfer, J. Wiley, N.Y., 1967, 348

[32] G.D. Parfitt (ed.), Dispersion of Powder in
      Liquids, Elsevier, Amsterdam, 1969

[33] L. Dintenfass, J. Oil Col. Chem. Assoc. 40
      (1957) 761

[34] L.O. Kornum, Rheol. Acta 18 (1979) 178

[35] S. Pila, Defazet 28 (1974) 314

[36] T. Ginsberg, FATIPEC XII (1974) 497

[37] F. Dobbels and J. Mewis, Chem. Eng. Sci. 33
      (1978) 493

[38] S.E. Orchard, Appl. Sci. Res(A) 11 (1962) 451

[39] W. Goring and N. Dingerdissen, Farbe u. Lack 83
      (1977) 270

[40] S.T. Harris, The Technology of Powder Coatings,
      Portcullis Press, London, 1976, 239

[41] J.C. Berg, A. Acrivos and M. Boudart, Advances
      in Chemical Engineering, ed. T.B. Drew,
      Vol. 6, 61, 1966

[42] C.C. Mill, Brit. Ink Maker 16 (1974) 172

[43] J. Mewis, Scientif. Report No. 4 IVP Lab.,
      Brussels (1970)

[44] R.B. McKay, Brit. Ink Maker 19 (1977) 59

[45] A. Doroszkowski and R. Lambourne, J. Colloid
      Interface Sci. 26 (1968) 214

[46] J. Mewis and R. de Bleyser, FATIPEC XI, Milano,
      Add. 9, (1972)

334

[47] M.W. Westgate, Off. Dig. Fed. Soc. Paint Technol.,
      32 (1960) 616

[48] A.N. McKelvie, Progr. Org. Coatings 6 (1978) 49

[49] K. Walters, Rheometry, Chapman and Hall, London,
      (1975) 56

[50] G. Van der Hoeden, Ind. Lack. Betrieb 30 (1962)
      425

[51] V. Zorll, Ind. Lack. Betrieb 33 (1965) 231

[52] ASTM D1545-63

[53] H.A. Gardner and G.G. Sward, Physical and
      Chemical Examination of Paints, Varnishes and
      Colors, Gardner Lab., Bethesda MD (1962)

[54] D.J. Doherty and R. Hurd, J. Oil Col. Chem.
      Assoc. 41 (1958) 42

[55] E.K. Fischer, J. Colloid Sci. 5 (1950) 271

[56] L.O. Kornum, FATIPEC XIV, ed. V. Takacs,
      Budapest (1978) 329

[57] J. Mewis, Chim. Peint. 33 (1970) 89

[58] A. Karnis, H.L. Goldsmith and S.G. Mason, J.
      Colloid Interface Sci. 22 (1966) 531

[59] D.C-H. Cheng and B.R. Parker, Proc. VIIth Intern.
      Congr. Rheol. C. Klason and J. Kubat (eds.)
      Gothenburg (1976) 518

[60] J.R. van Wazer, J.W. Lyons, K.Y. Kim and R.E.
      Colwell, Viscosity and Flow Measurement,
      Interscience, N.Y., 1963

[61] J. Mewis, J. Non-Newtonian Fluid Mechanics 6
      (1979) 1

[62] S. Middleman, The Flow of High Polymers, Inter-
      science, N.Y., 1968, 28

[63] R. Lapasin, A. Papo and G. Torriano, FATIPEC XIV, ed. V. Takacs, Budapest (1978) 653

[64] D.J. Highgate and R.W. Whorlow, in R.E. Wetton and R.W. Whorlow (Ed.), Polymer Systems: Deformation and Flow, Macmillan, London, 1968, p. 251

[65] P-J. Klijn, J. Ellenberger and J.M. Fortuin, Rheol. Acta 18 (1979) 303

[66] J.F. Hutton, Rheol. Acta 8 (1969) 54

[67] E. Husson, Dissertation K.U. Leuven, 1976

[68] H. Schmelzer, Farbe u. Lack 79 (1973) 11

[69] D. Hadjistamov and K. Degen, Rheol. Acta 18 (1979) 168

[70] D.C-H. Cheng, J. Appl. Phys. 17 (1966) 253

[71] K. Kuwano, J. Jap. Soc. Col. Matt. 50 (1977) 547

[72] T.A. Amfiteatrova, S.W. Jakubovich and L.V. Kozlov, Plaste Kautschuk 23 (1976) 610

[73] E. Dall'Orto, S. Vargin, S. Bollani and M. Pitzalis, Pitture Vernici 53 (1977) 3

[74] C.F. Goodeve and G.W. Whitfield, Trans. Faraday Soc. 34 (1938) 511

[75] K.M. Oesterle, Farbe u. Lack 80 (1974) 633

[76] K.M. Oesterle, J. Oil Col. Chem. Assoc. 51 (1968) 1007

[77] H. Green and R.N. Weltmann, Ind. Eng. Chem. Anal. Ed. 15 (1943) 15

[78] D.C-H. Cheng, Nature 216 (1967) 1099

[79] H.J. Freier, Rheol. Acta 3 (1964) 254

336

[80]   ASTM D2196-68

[81]   J. Pawlowski, Chem. Ing. Techn. $\underline{28}$ (1956) 791

[82]   E. Mylius and E.O. Reher, Plaste Kautschuk $\underline{19}$
        (1972) 420

[83]   H.A. Mercer and H.D. Weymann, Trans. Soc. Rheol.,
        $\underline{18}$ (1974) 199

[84]   P.F.S. Cartwright, Brit. Ink Maker, $\underline{8}$ (1966) 83

[85]   N.E. Hudson, M.D. Bayliss and J. Ferguson,
        Rheol. Acta $\underline{17}$ (1978) 274

[86]   J. Pryce-Jones, J. Oil Col. Chem. Assoc. $\underline{26}$
        (1943) 3 ·

[87]   T.C. Patton, J. Paint Techn. $\underline{38}$ (1967) 656

[88]   R.W. Kreider, Off. Dig. Fed. Soc. Paint Technol.,
        $\underline{36}$ (1964) 1244

[89]   P.E. Pierce and V.A. Donegan, J. Paint Techn.
        $\underline{30}$, (1966) 1

[90]   H.L. Beefermann and D.A. Bergen, J. Paint Techn.
        $\underline{38}$(492) (1966) 9

[91]   G.D. Shettye, Paint India (1973) 37

[92]   J. Mewis, Technical Note No. 1, IVP Laboratory,
        Brussels (1968)

[93]   W.K. Asbeck, D.D. Laiderman and M. Van Loo,
        J. Coll. Sci., $\underline{7}$ (1952) 306

[94]   P. Fink-Jensen, Färg och Lack $\underline{9}$ (1964) 1.

[95]   D.M. Gans, J. Paint Techn. $\underline{44}$ (1972) 68

[96]   G. Langer and U. Werner, Rheol. Acta. $\underline{14}$ (1975)
        237

[97]   K. Lederer and J. Schurz, Rheol. Acta 14 (1975)
       252

[98]   K. Walters and N.D. Waters, Brit. J. Applied
       Phys. 15 (1964) 989

[99]   N.D. Waters and M.J. King, J. Appl. Pol. Sci.
       17 (1973) 3069

[100] R.W. Williams, Rheol. Acta 18 (1979) 345

[101] N.I. Gaynes, Testing of organic coatings, Noyes
       Data Corp. (1977) 183

[102] R.H. Rowland, W.R. Stimson and N.B. Garlock,
       Nat. Paint Varnish Lacquer Assoc. Sci. Circ.
       No. 783, 1959

[103] H.J. Freier, Farbe u. Lack 69 (1963) 87

[104] N. Street, J. Oil Col. Chem. Assoc. 39 (1956)
       391

[105] M. Camina, Defazet 28 (1974) 117

[106] A.G. Epprecht, Farbe u. Lack 84 (1978) 86

[107] H. Bruss, Farbe u. Lack 85 (1979) 86

[108] Proces Viscometers, Contraves AG, Zürich.

[109] Anon. (1974) Australas. Printer 25(8), 11

[110] A. Voet, Ink and Paper in the Printing Process,
       Interscience, N.Y., 1952, 70

[111] W. Roth and S.R. Rich, J. Appl. Phys. 24 (1953)
       940

[112] H.E. Weisberg, Off. Dig. Fed. Soc. Paint Technol.
       36 (1964) 15

[113] R.R. Myer, Off. Dig. Fed. Soc. Paint Technol. 33
       (1961) 940

338

[114] J. Mewis, FATIPEC IX, Brussels, 3-120 (1968)

[115] M.B. Roller and J.K. Gillman, J. Coating
       Techn. 50 (1978) 57

[116] R.R. Myers, C.J. Krauss and R.N. Schroff,
       Proc. Vth Intern. Congr. Rheol., ed.
       S. Onogi, Univ. Tokyo Press, Tokyo, Vol. 1
       (1969) 473

[117] T.A. Turner, Brit. Ink Maker 15 (1973) 59

[118] C.C. Mill, J. Oil Col. Chem. Assoc. 51 (1968)
       861

[119] J.H. Taylor and S.L. Cozzens, Natl. Printing
       Ink Res. Inst., Bull. No. 53, 1959

[120] H. Strasburger, J. Coll. Sci. 13 (1958) 218

[121] V. Kelhä, M. Manninen and P. Oittinen, TAPPI
       57(4) (1974) 86

[122] C.C. Mill, J. Oil Col.Chem. Assoc., 50 (1967)
       396

[123] Anon. Publications JN11-74 and RA22-73,
       European Committee of Paint and Printing Ink
       Manufacturers' Association.

CHAPTER 7

INDUSTRIAL AQUEOUS SUSPENSIONS

By

K.M. Beazley

CONTENTS

1. INTRODUCTION

Particulate materials are frequently processed, transported and sold in the form of a suspension in a liquid, usually water. In the following we shall restrict ourselves almost wholly to aqueous suspensions; important non-aqueous suspensions such as paints and printing inks have been dealt with in Chapter 6. Suspensions form an important group of materials and their rheological properties are important, both from the point of view of their actual behaviour during the flow process to which they may be subjected, and as a monitor of the physico-chemical state of the suspended particles.

The addition of particulate material to a flowing liquid distorts the streamlines of that flow; this process requires energy and that expenditure of energy is manifested as an increase in the viscosity of the system. Many authors have used theoretical or semi-theoretical treatments to relate the viscosity of a suspension to the volume concentration of the suspended phase [1-13]. Such equations, however, describe only the viscosity – concentration relationships and say nothing about the viscosity – shear rate relationships; that is, they necessarily treat the suspensions as Newtonian fluids, when they may, in fact, show marked non-Newtonian behaviour.

This philosophy of treating non-Newtonian materials as Newtonian, and controlling industrial suspensions by assessment of behaviour in simple shear flow (often by a one point measurement) is almost universal, the justification being that it seems to work. The quali- fication can be made, since it will often be found that a change in formulation brings with it a need to change the level of shear viscosity at which control is exercised – tacit confirmation that the process "sees" a more complex rheology than the controlling rheometer. For instance, simple viscometric control is used in the process of paper coating, where a sus- pension of pigment and adhesive is applied in a layer a few micrometres thick to a paper web by a machine operating at high speeds. Several methods can be used for application, but the commonest one uses a flexible blade for the application. Blade coaters themselves fall into several categories, but it will be sufficient to consider the Rice-Barton or pond-type coater, shown in Figure 1.

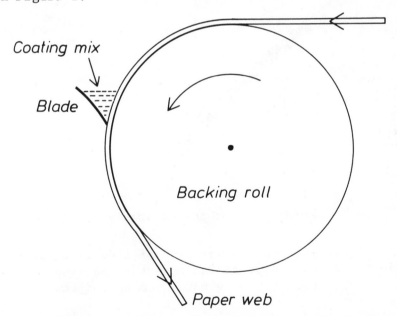

Coating mix

Blade

Backing roll

Paper web

Fig.1 Sketch of set up of pond-type paper coater.

The paper web is carried round a rotating, rubber covered backing roll; a flexible blade is pressed against the paper, and coating mix (commonly called coating "colour") held between blade and paper and doctored onto the paper as it passes beneath the blade nip. Figure 2 sketches the flow regime as the colour enters the nip. Circulation of colour in the pond will cause an expansion of the streamlines. At the entrance to the nip, there will be convergence, and hence elongational flow. In the nip, the colour will be subjected to shear rates in excess of 10^6 s^{-1} for a few microseconds while one can visualise that somewhere in the pond fracture must occur. Yet this rheologically complex process is controlled by a simple low shear viscometer.

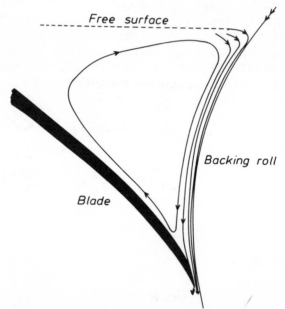

Free surface

Backing roll

Blade

Fig.2 Flow pattern in pond of pond-type paper coater.

Almost every aqueous suspension of industrial significance shows non-Newtonian behaviour in simple shear flow, which may originate from any of the following:

(a) forces of attraction or repulsion existing between the particles, which are modified by the applied flow. Figure 3 is a flow curve of an acid flocculated kaolin clay suspension (see section 2.1). The physico-chemical conditions are such that the material exhibits a yield stress, and behaves approximately as a Bingham plastic with time dependent (thixotropic) properties. The behaviour originates in the overcoming of the forces of attraction present between the particles in that state.

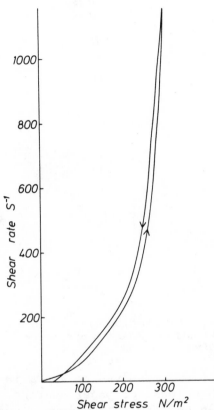

Fig.3 Flow curve of acid flocculated paper coating clay. 43% solids, Ferranti - Shirley.

(b)   attachment of particles to each other by polymer molecules (see section 2.5). This can be illustrated by the flow curve of Figure 4, which represents the behaviour of a dispersed kaolin suspension as such, and a similar suspension to which a small amount of a high molecular weight polymeric flocculant has been added.

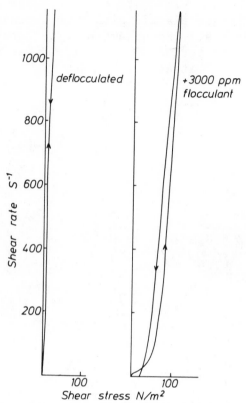

Fig.4 *Deflocculated paper coating grade clay,
without and with 3000 ppm of polymeric
flocculant. 40 % w/w solids,
Ferranti - Shirley flow curves.*

(c) the presence of dissolved polymers in the liquid phase. In the case of paper coating, the colour adhesive can be a soluble polymer, such as starch, and the elastico-viscous nature of the coating colour has been demonstrated from normal stress measurement with a Weissenberg rheogoniometer [14].

(d) the adsorption of the suspending liquid onto the particle surface. A deflocculated kaolin suspension invariably shows shear thinning behaviour at low shear rates (say, below 100 s^{-1}), even at shear viscosity levels of a few mPa.s; Figure 5 illustrates this.

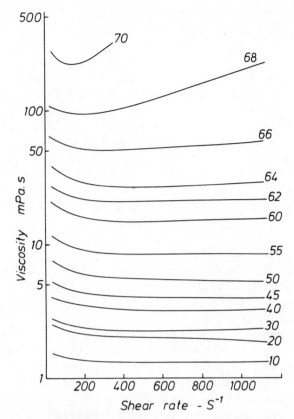

Fig.5 Viscosity vs. shear rate for deflocculated
paper coating grade clay slurries, at the
weight percent solids shown.
Weissenberg rheogoniometer data.

Since deflocculated particles repel each other, interparticle forces would give no immediate explanation of
this behaviour, and it is supposed that it originates
in the progressive shearing off of layers of adsorbed
water from the particles, so reducing the apparent
volume concentration of the composite clay/water particle, with a consequent decrease in viscosity.

(e) mutual mechanical interference between particles
in suspension. In suspension, particles will undergo
both translational and rotational motion. In the case
of an anisometric particle (e.g. laminar or acicular)
this rotational motion can result in entanglement, and
interference in mobility of one particle by another.
Such interference will increase with increasing shear

rate, with increasing solids content of the suspension and with increasing particle anisometry; the rheological manifestation of this interference will be a shear thickening behaviour. Figures 6a and 6b compare the behaviour for two kaolin clays. In the one case the clay is very fine (about 50% of the particles are finer than $\frac{1}{4}$ micrometre) and "chunky" (diameter to thickness ratio about 5). The particles pack well in suspension, and at the lower shear rates the suspension is quite mobile, but when shear thickening does occur, it does so severely. The suspension of Figure 6b is that of a near-monodisperse fraction of a ceramic clay, diameter -thickness ratio of about 50. Since the material is near-monodisperse particle packing is poor, shear thickening behaviour readily observed, and that at quite low solids. Elastico-viscous behaviour has been established in shear thickening suspensions from measurements made with a Weissenberg rheogoniometer [15,16].

(f) mechanical fracture of particles by shear forces. Removal of small irregularities or protrusions from particles is possible and such removal by shearing forces will nearly always make the suspension more fluid, since there is generally room between larger particles for these smaller ones to pack. The net effect is to impart a degree of shear thinning behaviour to the suspension, and this is almost always observed when mechanical fracture of particles (confirmed by monitoring particle size distribution) occurs. Use is made of this property in dispersing pigment suspensions to high solids contents. This particular behaviour is, of course, irreversible.

To summarise the preceding, we may note that factors (a), (b), (c), (d) and (f) might be expected to impart shear thinning behaviour, while factor (e) could be expected to give rise to shear thickening behaviour. Factors (a) to (e) inclusive could provide a basis for elastico-viscous behaviour.

Time dependence of behaviour (usually thixotropic) is also common. Flocculation arising from attractive forces normally takes a finite time to re-establish

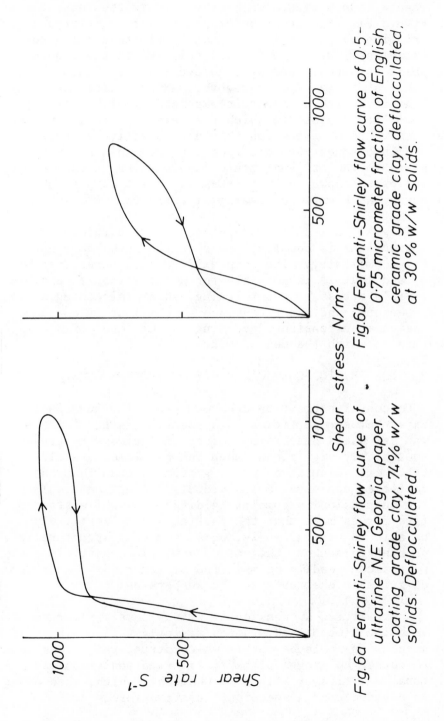

Fig.6a Ferranti-Shirley flow curve of ultrafine N.E. Georgia paper coating grade clay, 74% w/w solids. Deflocculated.

Fig.6b Ferranti-Shirley flow curve of 0·5–0·75 micrometer fraction of English ceramic grade clay, deflocculated, at 30% w/w solids.

Shear stress N/m²

Shear rate s⁻¹

itself when broken down by shear (note the loop of Figure 3). Flocculation originating from interparticle bridging by polymers also takes a finite time to re-establish itself. Dissolved polymers in the liquid phase may uncoil and stretch under shear - this, too, takes a finite time to recover. Shear thickening behaviour with anisometric particles involves taking them from a state in which they are comparatively free to move to one in which they are effectively "tangled up" into a near-solid; recovery from this near-solid state to the original fluid one is never instantaneous, and often shows up as a crossover point on a flow curve, as Figure 6a clearly illustrates.

It will be readily understood that the rheology of suspensions is complex, and research work bears out this complexity. The complexity can be underlined if we add that it is possible for a suspension to exhibit a yield stress, shear thinning, shear thickening, time dependence and elastico-viscous behaviour in one and the same suspension, depending on the level of shear rate at which the measurement is made.

2. THE PHYSICO-CHEMICAL BASIS OF SUSPENSION RHEOLOGY

Forces of attraction and repulsion will invariably exist between particles in suspension, since London-van der Waals dispersion forces will always be present, though only significant when the particles are close together. Repulsive forces generally mean that the particles can move as individuals (no aggregation) and the suspension is completely dispersed, while attractive forces mean that the particles are less likely to be able to move as individuals, and that structure is present throughout the suspension. Interparticle forces can readily be modified or reversed by addition of suitable chemicals to the suspension.

A structured suspension is often loosely referred to as being flocculated, though strictly a flocculated suspension would be one in which structural forces tend to gather up groups of particles into individual structural entities or "flocs". Chemicals which, when added to a structured suspension, initiate forces of repulsion between particles and thereby form a dispersed system are known as "deflocculants" or "peptising agents". The terms "dispersion" and "deflocculation" are often loosely used as being interchangeable. This is not strictly true since deflocculation refers to the establishment of repulsive forces between particles, while dispersion refers to their ability to exist and move as individuals; it is possible for a suspension of dispersed particles to exist where forces of repulsion are so small as to be insignificant.

The state of flocculation or deflocculation of the particles in a suspension is probably one of the most significant factors in determining its rheology, and we shall deal with it in some detail. The description of the mechanisms involved can best be illustrated by referring to the behaviour obtained with a specific material; for this purpose the behaviour of the mineral kaolinite (the main constituent of kaolin clay) will be used. Similar reasoning can be applied to other particulate material in suspension, though differences in response from material to material must naturally exist. For instance, one could hardly apply the concept of "acid flocculation" to a calcium carbonate suspension, while talc, possessing both hydrophyllic and hydrophobic surfaces [17] must have a surface chemistry differing in certain respects from that of kaolinite.

Kaolinite crystals occur in the form of thin hexagonal platelets a micrometre or so in diameter (Figure 7). Structurally the mineral is made up of alternating sheets of silicon and oxygen atoms arranged in tetrahedral coordination, and aluminium and hydroxyl ions in octahedral coordination [18,19]. A "slice" taken across a crystal would show successive layers:

Hydroxyl   (OH) ions
Aluminium   (Al) ions

Mixed oxygen and hydroxyl ions, where the layers "fit into" each other:

Silicon   (Si) ions
Oxygen   (O) ions.

Fig. 7   Kaolinite crystals

The structure is then repeated, starting with another layer of OH ions; hydrogen bonding holds the O-layer of one unit cell to the OH-layer of the next. What happens now when we place such a surface in water and change the pH? Three possible reactions could occur:

(i) $(Si - OH) + OH^- \rightarrow (Si - O^-) + H_2O$
      (The silicon layer, with the pH raised, becomes negatively charged).

(ii) $(Al - OH) + H^+ \rightarrow Al - OH_2^+$
      (The aluminium layer with the pH lowered, becomes positively charged).

(iii) $(Al - OH) + OH^- \rightarrow AlO^- + H_2O$
      (The aluminium layer, with the pH raised, becomes negatively charged).

In practice, a layer of silicic acid, H_2SiO_4 tends to be adsorbed on both basal surfaces, giving a surface whose reaction is similar to (i) above. For all practical purposes, then, we can view the basal (flat) surfaces of a kaolinite crystal as being negatively charged in water.

The crystal edges, however, have both exposed (Si - OH) groups and (Al - OH) groups present, and according to reactions (i) to (iii) the edge charge could be positive or negative, depending on pH. The point at which the sign of the charge changes with pH is just above neutrality for most kaolinites.

Suppose, however, that a particle were placed in water at such a pH that both faces and edges were negatively charged. Since the system as a whole must remain electrically neutral, redistribution of charges around the particle must occur to maintain neutrality. The origin of those redistributed charges will depend on what ionic materials happen to be present in the water at the time. Four cases of significance can be distinguished, plus one other which we shall briefly consider.

2.1 Acid Flocculation

Here, at acid pH's, the crystal is present with a positive edge charge and a negative face charge - a situation which has been confirmed by electron microscope studies of kaolinite particles onto which metal sols of specific positive or negative charge have been adsorbed [20,21,22]. Exchangeable cations present will tend to swarm around the crystal face, exchange-anions around the edge, producing "double layers" with the ion density decreasing with the distance from the particle (Figure 8).

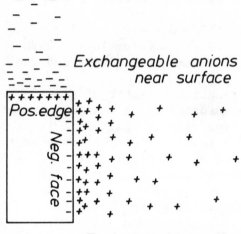

Kaolinite particle
seen edge-on

Fig.8 Sketch of distribution of charges in double layer around a Kaolinite particle.

The positive charge of one layer will attract the negative charge of another forming an "edge to face" flocculation and a "house of cards" structure through the suspension [23,24]. Microrheological considerations will therefore identify an acid flocculated suspension as being shear thinning and possessing a yield stress, as Figure 3 illustrated. The rheometry of flocculated suspensions has been comprehensively dealt with by Michaels and Bolger [24]. Figure 9 shows how the rheometry of acid flocculated kaolin suspensions varies as the suspension concentration increases.

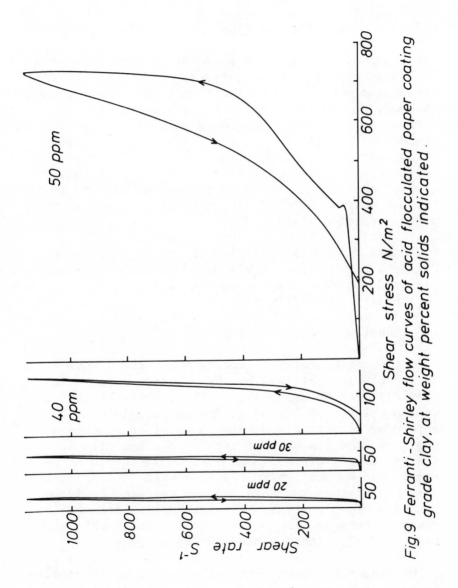

Fig.9 Ferranti-Shirley flow curves of acid flocculated paper coating grade clay, at weight percent solids indicated.

Shear thinning behaviour is evident and the presence of a yield stress not unreasonable; additionally, at higher solids the flow curves take the form of hysteresis loops of the type associated with thixotropic behaviour.

2.2 Alkaline deflocculation

We have said that above pH 7 the edge charges on a kaolinite crystal in water reverse sign, and the particle as a whole is negatively charged. Alkali, such as NaOH, have to be added to achieve this, and the cations of the alkali now contribute significantly to the double layer. This double layer is now composed throughout of an excess of cations swarming round a negatively charged particle. Interaction of the double layers now results in repulsion, with particles free to move as individuals, and the suspension is in a deflocculated state.

The state of deflocculation is comparatively weak. Figure 10 gives flow curves of an acid flocculated clay and the same clay at an alkaline pH. The increase in fluidity is readily seen.

2.3 Chemical deflocculation

Of greater commercial importance is the stronger deflocculation obtained by the addition of certain chemicals to the suspension. These materials, such as sodium polyphosphates or sodium polyacrylates (low molecular weight) dissociate in water to produce a short chain anion. The anion is absorbed strongly onto the positively charged edge – strongly enough to be considered as chemically reacting with it – and not only neutralises the edge charge, but also produces marked charge reversal. The interaction between the double layers so produced is much stronger than with simple pH adjustment alone [25-29].

The viscometry of a chemically deflocculated kaolin suspension is complex, particularly at the highest solids [15]. If a flow curve is plotted, four regimes can be found to merge one into the other, and these

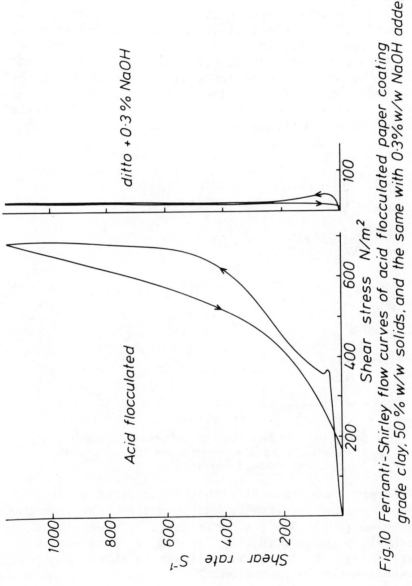

Fig.10 Ferranti-Shirley flow curves of acid flocculated paper coating grade clay, 50 % w/w solids, and the same with 0.3%w/w NaOH added.

will now be briefly described.

At low shear rates (below about 100 s^{-1} say), shear thinning behaviour can be observed; this was illustrated in Figure 5 and Figure 11 shows how that behaviour persists down to quite low solids contents. By definition, forces of repulsion exist between the particles; this, in itself, provides no explanation of the behaviour. It is reasonable to suppose that the ions in the double layer will tend to hinder the movement of water molecules in their vicinity and give rise to a layer of water, which, for want of a better word, we can describe as "semi-rigid".

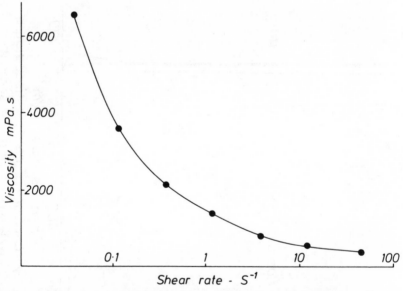

Fig.11 *Shear thinning behaviour of deflocculated paper coating grade clay at 68% w/w solids, at low shear rate. Weissenberg rheogoniometer data.*

Several authors deal with this adsorbed water, and quote values for its thickness varying from 5 nm to 3300 nm [30,31,32]. A basis for this shear thinning behaviour could then originate in the progressive stripping off of this water layer by shear, as indicated in the introduction.

This hypothesis is difficult to prove, but the influence of adsorbed water is usefully demonstrated by

Figure 12, which represents a flow curve of a fully deflocculated suspension of a ceramic grade clay. In
this instance the clay contains several percent of the
clay mineral montmorillonite; this mineral has a surface area many times greater than that of kaolinite
and its ability to form structure with water is demonstrated by the fact that a 0.5% suspension of it in
water can form a gel. The flow curve of this ceramic
clay is a hysteresis loop typical of a thixotropic
material. If the $-\frac{1}{4}$ micrometre fraction (containing
the montmorillonite) is removed from the clay by
centrifuging the structure is destroyed and a fluid,
virtually Newtonian suspension results.

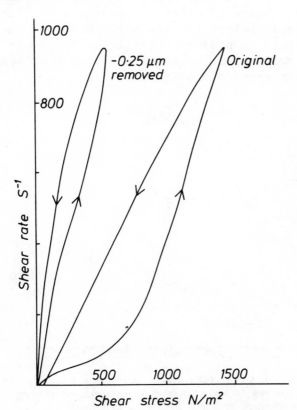

Fig.12 Deflocculated ceramic grade clay, at
65% w/w solids, as received and with
the - ¼ micrometer fraction removed.
Ferranti-Shirley flow curves.

From, say 100 s^{-1} to 1000 s^{-1} shear thickening behaviour is apparent, the degree of shear thickening
increasing with increasing solids content. Figure 13,
which illustrates this point, represents a series of
flow curves taken at increasing solids content. As
noted in 1(e) particle shape is a major factor in controlling the level of this shear thickening behaviour,
the solids content at which it can first be detected
viscometrically and the range of solids contents over
which it can be detected. A suspension of "chunky"
particles which pack well, such as calcium carbonate
or certain types of kaolin, do not show shear thickening behaviour until the suspension solids content is
quite high, normally a few percent below the maximum
attainable. Onset of shear thickening behaviour can
be sudden and violent (Figure 6a). More anisometric
particles tend to produce shear thickening more readily,
do so at lower solids contents and show that behaviour
over a wider range of solids contents.

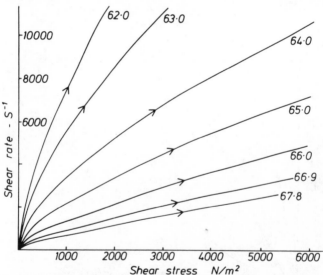

Fig.13 Ferranti-Shirley flow curves of deflocculated paper coating
grade clay, at weight percent solids indicated, showing
development of shear thickening behaviour.

Now, it can be shown optically [33] that clay particles in suspension rotate when that suspension is
sheared and it is reasonable to suppose that this
shear thickening is due to progress interference

between rotating particles as the shear rate is increased. Several authors [34-38] have accepted this mechanism, while a similar mechanism [39] has been used to describe certain facets of the rheological behaviour of suspensions of attapulgite, an acicular clay mineral.

The next regime which can be observed on a flow curve is a second region of shear thinning. It occurs at shear rates of a few thousand s^{-1}, and is illustrated in Figure 14. To produce such thinning something has to "give way"; in this case, fracture of the clay particles seems likely, as noted in 1(f).

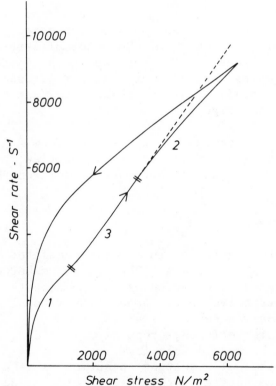

Fig. 14 Ferranti-Shirley flow curve of paper coating grade clay at 69% w/w solids, showing two shear thickening regions (1 and 2) and the second shear thinning region (3).

Finally, a second shear thickening region can be found, also illustrated in Figure 14. It will become

apparent at a few ks^{-1}. Its microrheological origin
is not well defined, but it is suggested that shear
forces pack particles into aggregates, and shear
thickening occurs from interference between these
aggregates - we might reasonably speak of dilatancy
in these circumstances, since roughly spherical
entities are involved. One simple experimental
observation exists which favours this suggestion.
When a flow curve showing two shear thickening
regions has been obtained on a Ferranti-Shirley cone-
and-plate viscometer, visible "grains" are sometimes
subsequently seen when the cone/plate gap has been
opened up for cleaning.

The presence of normal stresses in a concentrated,
deflocculated clay suspension has been noted, and
Figure 15 illustrates them. Weak normal stresses
appear at low shear rates (corresponding to the first
shear thickening region), while larger stresses,
increasing rapidly with increasing rate, can be
measured at the higher shear rates.

It must be stressed that these four flow regimes do
not exist as well defined separate entities, bounded
by specific shear rates, but must merge from the one
into the other.

2.4 Salt flocculation

Consider two particles in the deflocculated state,
separated by a distance d. The net force between
these particles is composed of two components [40]:
  (1) an attractive force, proportional to d^{-7}, due
to the London-van der Waals attraction.
  (2) a repulsive force, proportional to d^{-2}, due to
interaction between double layers.
Generally, the net result of these forces is a repulsion, and the particles remain in a stable, deflocculated state. If we suppose the particles to be forced
closer together, this will become increasingly more
difficult to do until a point is reached at which the
van der Waals attraction predominates and the suspension rapidly flocculates. The energy for this could
come from vigorous mixing or from raising the temperature.

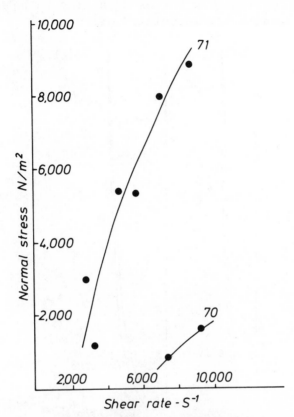

Fig. 15 *Normal stresses in deflocculated paper coating grade clay suspensions, at weight percent solids shown. Weissenberg rheogoniometer data.*

If we now add a soluble salt (e.g. NaCl) to the suspension, the electrochemical conditions around the particle are changed. Presence of the electrolyte increases the apparent dielectric constant of the liquid phase and reduces the repulsive interaction between the double layers until it finally disappears; particles then flocculate rapidly in face-to-face, edge-to-edge and edge-to-face modes. The flocs so produced are less bulky than "cardhouse" flocs and rheological structure in such suspensions is less marked.

Salt flocculation is of less practical importance than acid flocculation and polymer flocculation;

362

Figure 16 compares flow curves of a clay in both acid flocculated and salt flocculated states.

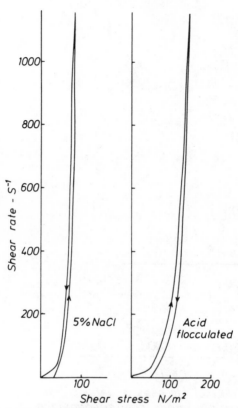

Fig.16   Ferranti - Shirley flow curves of acid flocculated paper coating grade clay at 40% w/w solids, and the same clay, fully deflocculated, and with 5% NaCl subsequently added.

2.5   Polymer flocculation

Polymer flocculation is widely practised in industries where suspended solids have to be removed from water. The clay industry itself tends to rely rather on acid flocculation and uses polymer flocculants only sparingly since they are strong flocculants whose effect is not always easily reversed.

The use of low molecular weight sodium polyacrylate as a deflocculant has already been noted; for this purpose its molecular weight might be a few thousand

and the mean length of its polymer chains 3 nm. It is
possible, however, to produce synthetic organic polymers with a molecular weight of 10^7 and chain lengths
of 20 micrometres. These chains are much larger than
kaolinite particles, and it is possible for one end
of such a chain to adsorb onto the surface of one particle and the other end to attach itself to another.
In this way low density flocs can be built up.

The rheology of suspensions of such flocs might
differ from that of acid flocculated suspensions since
there would now be polymer chains to be uncoiled.
Further, increasing flocculant dose would produce an
increasingly strong structure in the suspension, provided every available polymer chain could attach itself
to a particle. Acid flocculation is compared with
polymer flocculation at increasing doses of polymer in
Figure 17.

2.6 Dispersion with protective colloids

The old time paper coater who used casein as his adhesive would not normally use a chemical deflocculant
in the preparation of his coating mix since he knew
that the casein solution itself would produce an
acceptable state of dispersion in his clay suspension.
This it does by acting as a "protective colloid";
the technique is almost never used now in conjunction
with clay (casein is rarely used as an adhesive today)
but is of significance in the preparation of other
types of suspension.

In section 2.4 it was noted that, if two particles
could be brought close enough together to overcome
forces of repulsion between them and cross the energy
barrier appropriate to the situation, the particles
would flocculate. A protective colloid acts by
adsorption onto the particles, so forming a barrier
around them which makes it impossible for them to
approach close enough to flocculate. The particles
therefore remain in a dispersed state.

364

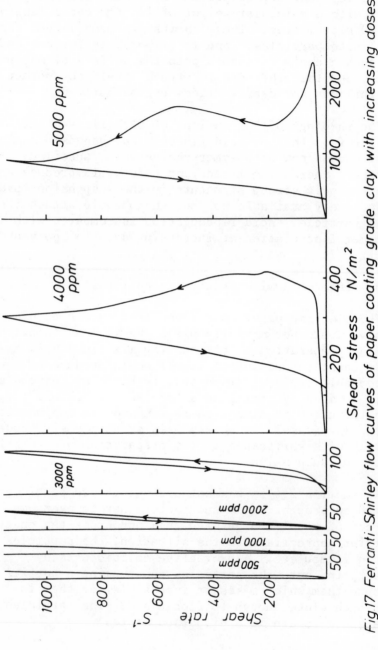

Fig 17 Ferranti-Shirley flow curves of paper coating grade clay with increasing doses of polymeric flocculant added. 40% w/w solids.

3. RHEOLOGICAL CONTROL IN VARIOUS INDUSTRIES USING SUSPENSIONS

The comments of the previous sections should have made it clear that the rheology of aqueous suspensions can be complex, and that the complexity is increased by the addition of materials other than pigment and water to the suspension, as would often be the case in practice. Industrial control, however, tends to ignore the complex rheology shown up by research, treats its materials as rheologically simple and uses simple - in some cases, primitive - methods for rheological control. This dichotomy between the research laboratory and shop floor must be stressed.

We shall consider the following industrial materials:

  (i)   Mineral pigments
 (ii)   Paper
(iii)   Ceramics
 (iv)   Cement pastes, concrete
  (v)   Drilling muds
 (vi)   Pharmaceuticals.

In each case we shall attempt to sketch what the basic rheology of the material involved might be and then contrast that with the methods of rheometry actually used by the industry concerned.

3.1 Mineral Pigments

Mineral pigments used in suspension could include kaolin clay (china clay), calcium carbonate (ground or precipitated), titanium dioxide, bentonitic (montmorillonitic) clays, calcium sulphate, barium sulphate, talc, coal, iron oxides, and others. Rheological control is not always exerted in the preparation of these materials, control being exerted rather in the finished materials made from them. The clay industry (and to a lesser extent the calcium carbonate industry) is the major exception here, and the tests used will be described in more detail.

3.1.1 Paper Clays

The paper industry currently consumes about 80% of high grade kaolin clay produced. A substantial proportion of this is used as filling for paper (i.e. included as a constituent of the paper sheet), but rheological properties are of little significance here. Rheological properties are of much greater importance in clays provided for paper coating, a process noted in the Introduction. The significance of rheological properties in the actual application of coating colour to the paper will be outlined later, but we may also note that it is of importance in the actual preparation of the colour; it is this aspect which is generally in mind in the rheological control of a clay intended for paper coating. Three points are of importance:

(i) The paper coater likes to have a pigment suspension at as high a solids content as possible, to reduce the amount of water that has to be dried from the coated sheet. To achieve these solids, the suspension has to be in a deflocculated state. The clay itself has to be as free from montmorillonitic clays as possible since these immobilise significant amounts of water and thereby reduce the attainable solids; thus, paper coating grades of clay are usually mined from deposits containing no montmorillonite or from those parts of deposits where the content is nil or minimal. The particle size distribution of a coating clay has to be such as to give good particle packing - a monodisperse clay packs badly and attainable solids with such a clay would be unacceptably low. Particle size distribution is controlled by processing.

(ii) Preparation of the pigment suspension will usually involve its dispersion in a high speed mixer. The presence of shear thinning behaviour in a chemically deflocculated clay has been noted (section 2.3) and if that shear thinning is excessive - due, say, to unsuspected incomplete deflocculation of the clay - then energy from the mixer will be wasted in overcoming that structure. Shear thinning behaviour is normally found in a high solids suspension; it is useful in that it ensures that particles hold each other firmly (to put it simply), so enabling the shearing mechanism of the mixer to break down aggregates and chip the rough edges from individual particles. The suspension viscosity is consequently reduced and an acceptable solids content attained. Excessive shear thickening, however, is undesirable because it may involve fracture of the suspension with the subsequent working of only a part of it by the mixer. What is actually excessive in this context depends on the clay (or other pigment) being made down, and the type and capacity of the mixer [41,42,43].

(iii) Screening and pumping are also involved. Here, too, excessive structure or shear thickening can cause a screen to block or a pump to operate unsatisfactorily.

Clay is being sold in increasing quantities as a high solids slurry, made down by the producer and shipped to the customer in that form. This form of shipment is also widely used for slurries of ground calcium carbonate destined for use in paper coating.

The remarks made here for clay would generally apply to other pigments used in coating, such as titanium dioxide, but these are normally comparatively minor constituents of a coating mix. Where significant rheological differences do exist they will be noted separately and contrasted to clay behaviour.

Low shear rate viscosity control

The Brookfield viscometer (Figure 18) is universally used. Results given by it are normally "one point" measurements, though as emphasised in section 2.3 clay suspensions are non-Newtonian fluids. The viscometer can, however, be run at four speeds (10, 20, 50 and 100 rpm) which may give a very rough idea of the non-Newtonian nature of the sample. The measuring head consists of a simple disc mounted on a spindle, and seven such spindles of different diameter are provided with the instrument. The disc is dipped into the liquid under test, rotated by a motor linked to it via a spring, and the torque registered by a pointer moving over a scale. A clutch mechanism disengages the pointer

Fig. 18 Brookfield viscometer

and clamps it to the scale so that the instrument can be stopped and read. Until recently [44] shear rate could not be precisely defined for a rotating disc, but an empirical treatment suggests [45] and experiment confirms [15] that the approximate shear rate in s^{-1} is equal to $\pi/10 \times$ rpm.

Low shear rheometry is used both to specify the viscosity level of an optimally deflocculated suspension at a given solids and to establish what that optimum deflocculant dose is. American and British practices in specifying slurry viscosity vary somewhat. American practice is to use the TAPPI Standard (T 648 - sm - 54) where the viscosity of a suspension at 71% W/W solids is measured at 100 rpm and quoted in mPa.s. English kaolin clays cannot always be made up at 71% solids and a different technique is used. It is found [46,47, 48] that over a restricted range of solids the reciprocal of the square root of the suspension viscosity (in poise) plots linearly against the weight concentration of solids. For the test, a suspension is prepared at as high a solids as possible, and viscosity measured at 100 rpm on the Brookfield. The sample is then diluted by 2% in solids and a second measurement taken. The root reciprocal viscosity is plotted against weight percentage solids and the solids required to give a viscosity of 5 poise (500 mPa.s) read off and quoted. This is called the "viscosity concentration" or "Clark viscosity".

Both types of measurement can be used for any suspension of pigment intended for paper coating. They are certainly used for calcium carbonate slurries.

Measurement of optimum deflocculant dose is carried out by measuring the change in viscosity as deflocculant is progressively added to a flocculated slurry. The viscosity will pass through a minimum, and the level of deflocculant required to achieve that minimum is taken as the optimum deflocculant dose [49]. The optimum deflocculant level and viscosity at that optimum vary with suspension pH, as Figure 19 illustrates. Recent work [50] indicates that the optimum depends also on the rate of shear at which the measurement is

370

made. Here, a high shear capillary rheometer was used both to measure viscosity *vs*. shear rate and viscosity *vs*. deflocculant dose of complete coating colours (clay, or calcium carbonate, plus adhesive) containing various levels of different types of deflocculant. The results showed that generally the optimum deflocculant dose increased with increase in shear rate. It should be stressed, however, that low shear rate measurements are customarily used for control purposes.

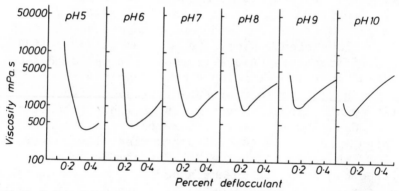

Fig.19 Deflocculation curves of paper coating grade clay with Sodium Polyacrylate, at pH levels indicated. Brookfield viscometer data, 10 r.p.m.

High shear rate measurements

Other than capillary rheometers, no commercial rheometer exists capable of measuring suspensions at the shear rate of 10^6 s^{-1} relevant to a paper coating machine [51], though band viscometers [52] are able to operate at 10^5 s^{-1}. A Brookfield viscometer, of course, cannot give information bearing any close resemblance to that obtaining at these shear rates. It is felt that commercial rheometers such as the Ferranti-Shirley cone-and-plate (Figure 20), or Hercules Hi-Shear (Figure 21) give a better indication of what might be expected under conditions of relatively high shear, even though the shear rates attainable

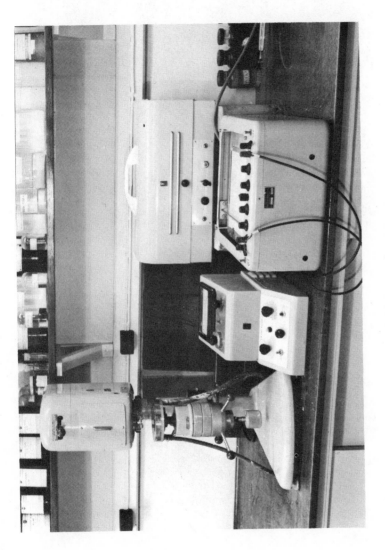

Fig. 20   Ferranti-Shirley viscometer

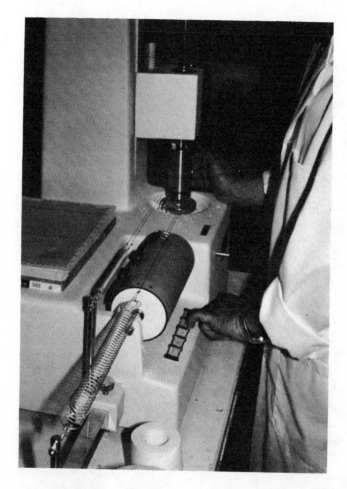

Fig. 21   Hercules viscometer

with them (up to 2×10^4 s^{-1}) are two orders of magnitude below that in the coating machine. The essential behaviour of a clay suspension up to about 10^4 s^{-1} has already been sketched (section 2.3). Flow curves obtained on kaolin suspensions with a capillary rheometer [50] could be interpreted as showing the same two shear thinning and two shear thickening regions indicated by rotational viscometry; additionally these measurements show the existence of a third shear thinning regime at shear rates in excess of 10^4 s^{-1}.

British control practice does not include a high shear control of slurry, although a very few paper mills use a high shear rheometer in their own mills. Though no official standard or published work exists to define it, American practice does include a high shear rate control, invariably made on a Hercules viscometer, which is a simple concentric cylinder instrument. A plot of shear rate *vs*. shear stress is recorded automatically by the instrument, presented as "Torque, dyne.cm $\times$ 10^5" *vs*. "Angular velocity, rpm", and clay viscosity will be specified by quoting the torque reading for a given rpm, usually the maximum. A clay would then be referred to as an "18 dyne clay" or a "12 dyne clay" or whatever it might be; the terminology is universally applied and understood. Figure 22 shows a typical Hercules recording for a clay slurry.

3.1.2 Other mineral pigments used in paper coating

Calcium carbonate, used in paper coating, is either a precipitated or ground, natural material. Precipitated carbonate is produced in either a calcitic form (particles roughly cubic) or aragonitic (acicular). Where the material is supplied as a dried powder, as is often the case, severe aggregation may be present (Figure 23). Slurries of this material are generally markedly shear thickening [53], so much so for it to be considered disadvantageous. Ground carbonates, on the other hand, are always calcitic, often supplied in slurry form, and hence do not suffer from this problem of aggregation [53,54]. Their particles are cubic and pack well. If this is linked to the material's low

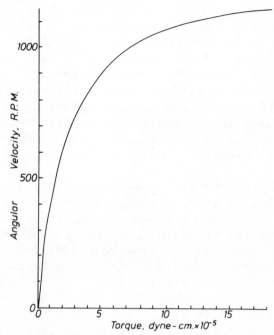

Fig.22 Hercules viscometer trace.

surface energy, implying that any adsorbed water layer
will be small compared with that of clay [55], it will
be seen that a ground carbonate whose particles pack
well, will give a suspension of higher solids content
than can be achieved with clay. One might therefore
expect that the structure apparent at low shear rates
with a clay suspension would be less, but as illustrated in Figure 24, structure is still significant.
Shear thickening behaviour has been demonstrated with
a Hercules viscometer on ground carbonate slurries at
70% W/W solids.

A little fine ground talc is used in coating. Like
calcium carbonate it has a low surface energy [55];
unlike carbonate, its particles are thin plates, with
diameter : thickness ratios which can be as high as 50
or 60. The particles are therefore mobile in deflocculated suspension, interfere readily with each other's
motion, and yet can readily be forced past each other
by shear forces. A flow curve would therefore show
significant shear thickening behaviour, but with a considerable degree of apparent structure [56]. The word
"apparent" is used since if the particles repel each

Fig. 23   Aggregation in Calcium carbonate

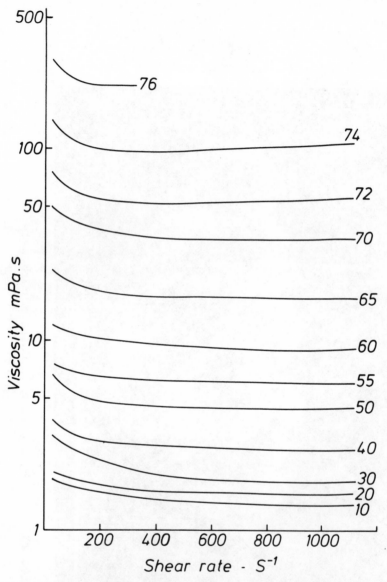

Fig. 24 Viscosity versus shear rate for
deflocculated ground calcium carbonate
slurries, at the weight percent solids shown.
Weissenberg rheogoniometer data.

other (as they should do in the deflocculated state)
and adsorption of water is minimal (as low surface
energy would imply) then there seems to be no immediate mechanism to hand to explain the presence of true
physico-chemical structure in the suspension.

3.1.3 Ceramic clays

Whereas the paper coater strives for a clay which
will make up to as high a solids content as possible
in the fully deflocculated state, and with as little
structure as possible, the ceramist will look for a
certain level of structure in his clay suspensions.
The mixtures of minerals and water used for the manufacture of ceramics ("bodies") are used either as a
solid, but mouldable plastic material, or as pourable
suspensions.

The former is used in plastic forming processes,
such as hand forming or pressing. The latter is utilised in the process of slip casting; here the suspension is poured into a plaster mould shaped in the
form of the article being made (a cup, say). Water is
drawn from the suspension into the mould by capillary
attraction and a solidified layer is built up in the
mould. When the layer is sufficiently thick,excess
suspension is poured from the mould and the adhering
layer left to consolidate. It can then be removed.

The mineralogical constituents other than clay have
little affinity for water. If a fully deflocculated
paper clay were used in the body the resultant plastic
body or cast would be flabby and possess little co-
hesion; in the case of a cast drainage would be poor
since the kaolinite particles would move freely,
orient and pack into a thin layer of parallel particles.
A ceramic clay is therefore required to impart a
certain degree of structure, both to assist drainage
in slip casting and to impart some rigidity to the
finished article. This structure will be achieved and
its extent controlled by using the body in a partially
deflocculated state.

Figure 12 illustrated the extent of structure, both
in terms of shear thinning and thixotropic behaviour,

that can be achieved with a ceramic grade of clay.
Whether or not a true yield stress is present in the
fully deflocculated state is debatable, since forces
of repulsion exist and there seems to be no physico-
chemical mechanism to account for a yield stress in
these circumstances [57,58]. Figure 25 illustrates
the flow behaviour of a casting slip at various states
of deflocculation [59], measurements being made on a
specially designed low shear rate viscometer. The
suspension is measured flocculated, under-flocculated,
fully deflocculated and over-deflocculated. The flocculated material (at a lower solids content than the
others) does show a positive yield stress and virtually no thixotropy; it approximates to a Bingham plastic. The under-deflocculated suspension also shows a
yield stress but shows marked thixotropy. The fully
deflocculated suspension is virtually Newtonian, while
the over-deflocculated material shows no yield stress,
is shear thinning with slight thixotropy, and is more
viscous than the fully deflocculated material.

The only rheological control, however, applied in
Britain to a casting clay is that of casting concentration [60]. Sodium silicate, the deflocculant normally used in the industry, is added to a clay suspension (or body suspension) and the viscosity change
followed with a Brookfield viscometer run at 20 rpm.
A slurry is then taken at that silicate dose giving
minimum viscosity (the optimum deflocculant dose) and
progressively diluted. Viscosity is measured at 20
rpm during dilution, and that concentration giving a
viscosity of 500 mPa.s is quoted as the casting
concentration.

3.2 Paper

Paper is customarily made by applying a thin layer
of a dilute suspension of cellulose fibres, plus other
materials, to an endless moving belt of wire mesh (a
Fourdrinier wire). Water is subsequently removed by
gravity and by vacuum applied to the wire, and by
pressure and heat subsequent to removing the paper
from the wire. That the flow of fibre suspensions is
non-Newtonian is apparent to anyone who has watched

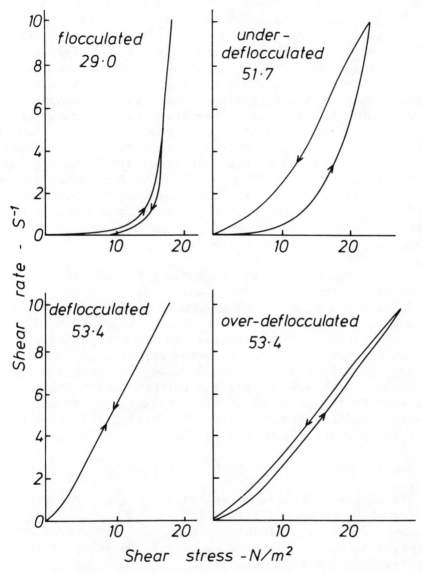

Fig. 25 Flow curves of ceramic slips at various
states of deflocculation. Specially designed
viscometer. Slips at volume concentration
solids indicated.

such a suspension circulating in a hollander beater and
seen the clear evidence of plug flow in it; a more
detailed study of fibre flow characteristics in a glass
tube [61] has established that here three regimes of
flow may exist at suspension concentrations of about
0.2% by weight. These regimes have been designated
plug, mixed and turbulent flow. The mixed flow region
represents a transition from plug to turbulent flow
near the wall and plug flow at the centre. Despite
this complex behaviour, fibre suspension rheology is
never measured or controlled in practice, other than in
a loose sense in the control of flow rates. The rheological properties of the made paper are of much greater importance, are more comprehensively measured and of
importance in controlling the treatment to which the
fibres are subjected during preparation and the chemical additions made to the paper, but are not relevant
in the context of suspensions.

Rheology, however, is of considerable significance in
the operation of paper coating, where a mixture of an
adhesive and pigment suspension is applied to the surface of a paper web. Points of importance in the
preparation of the suspension and the complex nature of
flow which may well occur in the application have already been considered (Introduction, sections 2.3 and
3.1.1). Here we consider the further complications
caused by the presence of the adhesive, since its presence modifies significantly the rheological properties
of the pigment suspension to which it is added.

Adhesives are of two basic types:

Soluble polymers, such as starch, casein, polyvinyl
alcohol or carboxymethyl cellulose (CMC)
Dispersions (lattices) or insoluble polymers or copolymers, such as styrene butadiene, polyvinyl acetate,
polymethylmethacrylate, etc.

The rheology of the coating mix depends to quite a
large extent on the type of adhesive used. If a mix is
made up consisting wholly of pigment and latex adhesive
there will normally be very little structure present.
The mix could be almost Newtonian and might splash or

mist to an undesirable extent during application. If
the solids level is raised to increase the viscosity,
the mix can be markedly shear thickening, since both
pigment and adhesive are effectively "solid" and interference between the constituents could be significant.
Figure 26 illustrates the transition from near-Newtonian
to shear-thickening behaviour experienced as mix solids
is increased. A shear-thickening mix could coat uneven-
ly, or extrude under the coating head in the form of
"stalactites". Further, the mix is in contact with the
paper for a finite time between application and doctor-
ing, during which time water will migrate from the mix
into the paper, so that the mix arrives at the doctor
at a higher solids than is present in the bulk mix [62-
68]. All-latex mixes are particularly prone to this
phenomenon, and an all-latex mix will normally contain
perhaps 1% of soluble polymer, which imparts both a
desirable level of shear thinning structure and controls migration, or the latex may itself contain a
soluble polymeric material which reacts to impart
structure when the mix pH is raised above neutrality.
The latter system has the advantage that the mix can
be prepared in a fairly fluid state, and so be easily
pumped and screened, with structure being imparted at
a fairly late stage in the preparation.

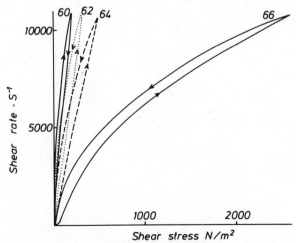

Fig.26 Ferranti-Shirley flow curves of paper coating
mixes, with SBR latex as the sole adhesive. Mixes
at the weight percent solids indicated.

The level of latex used will depend on the printing process to which the paper will be subjected, and may vary from 5% for a paper to be printed gravure to 12% for a paper to be printed offset.

Soluble polymers are now not often used as the sole adhesive; when they are, an offset coating might contain perhaps 15% starch, or 10 – 12% casein, or 5 – 6% polyvinyl alcohol. Rheologically, shear-thickening behaviour is now unlikely since the adhesive is in the form of a solution. The coating mix will now normally be structured, with consequent shear-thinning behaviour, and will exhibit thixotropy. The presence of the polymer can impart significant elastico-viscous behaviour; this has been demonstrated with a rheogoniometer [14] and a capillary rheometer [69,70]. An all-latex colour can also show elastico-viscous behaviour [71], due to the "solidity" imparted to it by the action of shear in a shear-thickening situation.

Figure 27 illustrates a flow curve for a starch-based coating mix, while Figure 28 demonstrates the presence of elastico-viscous behaviour in such a colour.

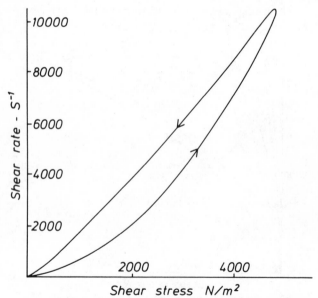

Fig.27 Ferranti-Shirley flow curve of paper coating mix with starch (100 parts pigment : 20 parts starch) as the sole adhesive.

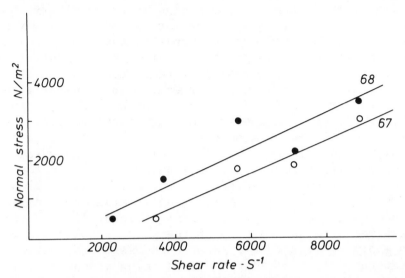

Fig. 28   Normal stresses for starch-bound paper coating mixes,
at the indicated weight percent solids.
Weissenberg rheogoniometer data.

The rheological treatment to which the mix would be
subjected in the common process of blade coating was
sketched in the Introduction.   The complex nature of
coating mix rheology has been underlined, while the
difficulties encountered in attempting to apply rheological concepts to paper coating are exemplified by
those studies of blade coating where various authors
attempt to relate the weight of coating applied (an
important parameter) to the coating rheology and
machine conditions [14,72-82].   Yet, as noted earlier,
the practical paper coater often controls his mix
rheology by a one point measurement on a Brookfield
viscometer, knowing from experience that if his mix
viscosity as measured lies between certain limits it
will (or should) coat satisfactorily, and that if his
mix formulation is changed he will probably have to
change his control limits.   According to the type of
mix being coated, which may vary for several reasons,
the mix Brookfield viscosity at 100 rpm could vary
from 1000 to 3000 mPa.s.

Some paper coaters do use a relatively high shear
rheometer, such as a Ferranti-Shirley or Hercules as a

control instrument, while at least one coating mill is
known by the author to use a high shear capillary rheometer. Such instruments, however "simple" they may be
to operate, are neither as simple to use as a Brookfield,
nor do they have the practical merit of being portable;
they are accordingly generally regarded as being in the
domain of the research department, should the mill have
one, rather than for day-to-day control. Coaters applying mix by roll coater, however, might still utilise
as a control factor a concept called "levelling index"
[83], which relates to the mix thixotropy and which
does require a relatively high shear rate rheometer for
its evaluation. The degree of thixotropy is of greater
significance in roll coating than in blade coating
since roll coating involves a significant film splitting mechanism at the nip of the pair of rollers used
to apply the mix to the paper. Excessive thixotropic
structure may mean that the patterning produced by the
film splitting has not had time to flow out before the
coating dries.

3.3 Ceramics

(a) Casting Slips. The non-Newtonian nature of casting slips has always been emphasised (section 3.1.3)
yet their rheological control in practice is extremely
simple. In Britain the Gallenkamp torsion viscometer
(Figure 29) is almost universally used [59,60]. A
small metal cylinder is suspended from a torsion wire
and dips into the suspension to be tested; the position and deflection of the measurement head is indicated
by a pointer and scale, graduated in degrees. In use,
the head is immersed in the sample, twisted through 360
degrees and released. The extent of "overswing" (past
the zero mark) is a measure of the fluidity of the suspension, the greater the overswing the more fluid the
suspension. In practice, two measurements would be
made. A "fluidity" measurement would first be made on
a freshly agitated sample. The sample would then be
allowed to stand for a given period and the thixotropy
determined by a second measurement on the suspension,
now possessing some degree of structural recovery. In
Europe, an efflux device of the Ford cup type, or a
simple cylinder with a hole in the bottom, might be
used to provide similar information.

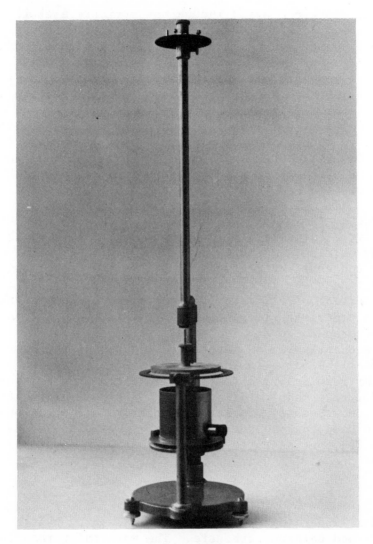

Fig. 29   Gallenkamp torsion viscometer

For trouble shooting, rotational instruments, such as the Brookfield or the Contraves Rheomat, are sometimes used.

(b) Plastic clays. Though not strictly a fluid suspension, a few words on this aspect of the industry ought to be said. The plasticity of a ceramic body intended for use in a plastic deformation process is almost impossible to define objectively, though experimental work on the deformation of plastic rods in torsional oscillation [84] demonstrated the presence of viscoelastic behaviour in such materials, and related that to the concept of plasticity and to accounting for the yield stress observed in them. Control in practice, however, is more often than not exerted by the experienced thumb of the potter.

One control that could be made is that of modulus of rupture, determined on a dry but unfired rod of the body or clay. Rupture is achieved in three point bending and could be measured on a home-made apparatus of the type shown in Figure 30 [60]. Here, the rod is supported on cylindrical supports and loaded centrally by a container slung across it. This container is counterbalanced to give zero initial load: water is subsequently poured into it. Breakage of the sample closes the valve in the water supply. The volume of water required to break the sample can then be converted into a modulus of rupture by a simple calculation. Various types of simple penetrometer have occasionally been used to assess the workability of plastic clays [85–88], but as noted above subjective methods are widespread.

3.4 Cement pastes, concrete

Portland cement is produced by calcining together clay and calcium carbonate. The material formed is a mixture of complex oxides, of which four can be noted:

Alite, $3CaO.SiO_2$, present as rectangular or
hexagonal crystals,
Belite, $2CaO.SiO_2$, present as small rounded grains,
Tricalcium aluminate, $3CaO.Al_2O_3$, of cubic habit,
Small amounts of tetracalcium alumino ferrite,
$4CaO.Al_2O_3.Fe_2O_3$.

Fig. 30   Rupture apparatus

Particle size would be expected to depend on the raw materials used and the conditions of calcining. These formulae are conventionally abbreviated to C_3S, C_2S, C_3A and C_4AF, using a single symbol for the appropriate oxide.

Unlike most industrial suspensions, the constitutents of cement react chemically with the water in which they are suspended; this, of course, is the basis on which cement works, chemical reactions with the water converting a paste which can be deformed into a solid material able to bear loads. A brief description of these hydration reactions will not be out of place [92].

$$C_3A + 6H_2O \rightarrow C_3A.6H_2O$$

$$2C_3S + 6H_2O \rightarrow C_3S_2.2H_2O + 3C.H_2O$$

These reactions are strongly exothermic and cause irreversible stiffening of the paste; in the case of the former, reaction is completed within a minute, which is undesirable. To slow down this reaction gypsum is ground into the cement when the calciner product (in lump form, or "clinker") is pulverised; the heat of the grinding mill converts the gypsum from a calcium sulphate dihydrate into a mixture of hemihydrate, $CaSO_4.\frac{1}{2}H_2O$ and anhydrite, $CaSO_4$. The latter is soluble, reacts with the $C_3A.6H_2O$ to produce ettringite. This material, present as acicular crystals, is formed on the surface of the hydrating C_3A, so providing a protective barrier against further rapid hydration.

C_2S hydrates more slowly, and is thought to contribute mainly to the long term development of strength in the set concrete. C_4AF reacts quickly along with the C_3A, both contributing little to the set strength.

The rheology of set cement, is beyond the scope of this chapter. It should be clear, however, that a suspension of particles which are rectangular or hexagonal, as C_3S would be, or acicular, as would be the ettringite or reprecipitated gypsum (which also forms during the period of hydration) and which also reacts chemically with the suspending phase, is likely to be markedly

non-Newtonian. This is, indeed, the case, and earlier
work with a concentric cylinder rheometer [93] showed
a cement paste to be a strongly shear thinning, thixotropic material, possessing a yield stress. The time
dependence of the shear stress, τ, was shown to obey
the relationship

$$\tau = \tau_o + (\tau_o - \tau_e) \exp(-Bt),$$

where τ_o was the initial torque, τ_e the final torque,
t was time and B a constant.

Later work with a modified Ferranti-Shirley viscometer [94,95] showed that a plot of $\ln(\tau_o - \tau_e)$ could
take one of three forms, illustrated in Figure 31, and
that the type could be altered progressively and re-
producibly from one to the other by any change that
could be expected to reduce the amount of hydration,
e.g. by increasing the water content of the paste,
decreasing the hydration time, decreasing the hydration
temperature or decreasing its specific surface.

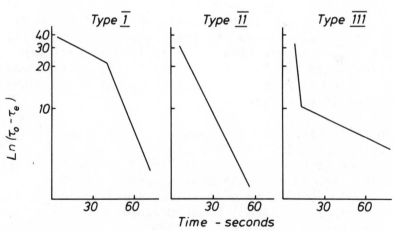

Fig.31 Ln $(\tau_o - \tau_e)$ plots versus time for cement pastes.

Later work with a rheogoniometer [96] demonstrated
that, although a cement paste exhibited a yield stress,
the subsequent behaviour was non-linear, and could be
either shear thinning or shear thickening, according
to the water-cement ratio and mixing time of the paste.
Lower rates and longer mixing times tended to be associated with shear thickening. The work also showed that
shear stress and shear rate q were related by a power
law equation of the type [97]

$$\tau = A(q + c)^B$$

where A and B have similar connotation to that found
in the traditional power law model, and c acts as a
correction term.

Extension of this work [98] demonstrated that a
water layer appeared on the surface of a cement paste
under shear; the results also suggested that, for a
cement paste in cone-and-plate flow, there existed a
critical shear rate below which the paste behaved as
an elastic solid, and slip occurred, and above which
behaved as a fluid.

Despite the clearly complex rheological properties
of a cement paste, instrumental control of these materials is not always used in practice, and when it is,
is of a very simple kind. The appropriate British
Standards (BS 12, "Portland Cement") may be followed:
the instrument prescribed by this is the Vicat penetrometer (Figure 32).

In this instrument a capped rod can move vertically,
carrying a pointer over a scale as it does so. At its
lower end it carries a "needle" used to penetrate the
sample; the mass of the moving assembly is 0.3 kg.
The sample is held in a stainless steel mould resting
on a non-porous plate. Three measurements are
specified:

Initial setting time - Here the penetrating element
is a needle, cross sectional area 1 mm^2. The needle
is lowered gently onto the sample of cement paste and
allowed to sink into it. The process is repeated until
the needle, when brought into contact with the sample,

Fig. 32   Vicat penetrometer

does not pierce it completely. The time between addition of water to the cement and the stage where the
needle ceases to pierce the sample is defined as the
initial setting time.

Final setting time – A needle provided with a 5 mm
diameter annular attachment is used. The cement is
considered as finally set when, on applying the needle
to the sample, the needle makes an impression on it
while the attachment fails to do so.

Normal consistency – The needle is replaced by a
plunger, 10 mm diameter. The quantity of water required to produce a paste of normal consistency is
0.78 of that required to give a paste which allows
settlement of the plunger to a point 5 to 7 mm from
the bottom of the mould when the paste is tested.

The standard also gives details of the procedures to
be used when loading the paste into the mould.

This, then, is the one simple instrumental test
specified for the rheology of cement paste, yet in
practice even this would probably be ignored, and the
cement tested by incorporation into a concrete and one
of the even simpler tests specified for concrete in
BS 1881, 1970, "Methods for testing concrete. Part 2"
would be used. The most widely used of these is probably the slump test. In this, a truncated cone, 8"
diameter at the base, 4" at the apex and 12" high, is
placed on a board with the large end downwards and
filled with concrete. The concrete is tamped in place
with a rod as it is added. The cone is then lifted
off vertically – the decrease in height of the concrete
on the board is a measure of the slump.

Allowable slump levels for concrete used in civil
engineering practice vary according to end use.

Concrete itself is effectively a suspension of fairly
coarse mineral particles (aggregate) in a cement paste.
The rheology of the paste itself is probably the controlling factor of the rheology of the composite system,
though unpublished work [99] suggests that with

increasing content of aggregate a fairly well defined
boundary solids exists at which the overall rheological
characteristic changed from shear thinning to shear
thickening.

3.5 Drilling muds

When drilling holes through rock strata, liquid
(drilling mud) is circulated between the surface and
the bore hole. This mud will serve three basic
purposes:

(i) It has to lubricate the bit. A Newtonian fluid
would serve this purpose, but the constraints imposed
by the other purposes demand that a shear thinning
fluid be used. The rate of shear thinning and ultimate
viscosity must be such that a material of the correct
viscosity is presented to the bit, "correct" being that
which is appropriate to the type of bit and conditions
of drilling.

(ii) When drilling through porous rock, such as sand-
stone, fluid mud will readily penetrate that rock.
Recovery from shear thinning behaviour should be rapid
enough to ensure that the mud gels, and effectively
seals the rock around the bore.

(iii) The mud must be able to act as a suspending
fluid, carrying the drill chippings to the surface.
Recovery from the shear thinning occurring at the bit
should ensure that the mud viscosity recovers to a
sufficient value to ensure that settling is minimal.

Considerations of bore geometry suggest that a wide
variety of flow geometries could occur during drilling
– laminar, elongational, helical, turbulent would all
be possible.

A practical mud probably contains four basic ingredients – water (the basis of the suspension), a clay
such as sodium bentonite, a soluble polymer and fine
ground barytes.

Sodium bentonite is composed of the clay mineral
montmorillonite, which has a characteristic layer

structure [100-103]. The large surface area of montmorillonite and its ability to form a gel in water was
noted in the earlier sections dealing with kaolin clays.
This large surface area and gelling behaviour in water
derive from its crystallographic and ionic structure.
Montmorillonite is composed of tetrahedral and octahedral layers similar to those of kaolinite, but their
arrangement differs. In kaolinite, the tetrahedral
and octahedral layers are stacked alternatively, like
a series of open Danish sandwiches. In montmorillonite
each repeat unit consists of two silicon-oxygen tetrahedral layers with the apices pointing towards each
other and enclosing an octahedral layer, rather like
the traditional English sandwich. Further, in montmorillonite, these structural layers are disposed with
their a and b axes in random orientation [102]. Montmorillonite has the ability to pick up cations between
its structural layers; certain cations (such as
sodium) also enable it to take up water between these
layers, causing it to swell. This water can be adsorbed from either the gaseous or liquid phase, but in
the case of liquid water sufficient can be taken up to
cause the structural layers to separate, and for the
suspension to form a quite strong gel. A flow curve
of a 6% W/W suspension of a sodium bentonite illustrates the degree of shear thinning behaviour which
this gel exhibits (Figure 33).

Soluble salts picked up as the bit passes through
certain strata can influence the rheological behaviour
of the bentonite. Calcium ions in particular, are a
nuisance in that they cause the suspension to gel too
severely; their influence is overcome by adding
soluble lignosulphonates to the mud.

The role of the soluble polymer in the mud is to
further impart shear-thinning structure and enable a
certain degree of structural control to be achieved.
Polymers used may include carboxymethyl cellulose,
guar gum, hydroxyethyl cellulose, all of which are
straight chain polymers, and xanthan gum, a "hedgehog"
or globular polymer. These, by virtue of the fact
that they are readily hydrated and possess chain segments which distort under shear, form shear thinning

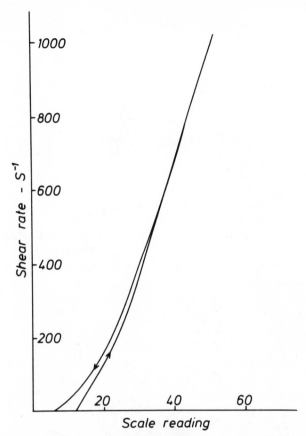

Fig.33 Fann viscometer flow curve for 6%
Bentonite suspension.

solutions in water, as Figure 34 illustrates. The
rheology is naturally strongly dependent on polymer
molecular weight and concentration. Carboxymethyl
cellulose is also known to possess the ability to de-
flocculate mineral suspension [104] and one cannot
discount the fact that the rheology of the drilling
mud may also be influenced by any physico-chemical
effect which the polymer may have on the mineral
content.

Like bentonite, polymers may also be influenced by
soluble salts encountered during drilling; in this
case the polymer itself is changed, probably for a
non-ionic and less readily affected one.

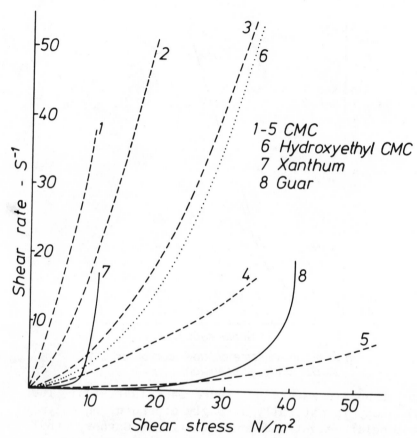

Fig.34 Flow curves for various commercial
Carboxymethyl cellulose solutions, and
solutions of similar cellulose derivatives;
with comparative data for Guar and
Xanthum gums.

The straight chain polymers noted above might also be expected to impart some degree of elastico-viscous behaviour to the mud, in the same way as they do to a paper coating mix.

Barytes has a specific gravity of 4.5 and is added to the mud to increase its density, since a suspended particle will not settle in a liquid of the same density, and the closer the mud density is to that of the drill chippings, the less their tendency to settle. The presence of barytes is unlikely to influence the non-Newtonian characteristics of the mud, though it would increase its apparent viscosity by a simple concentration effect.

Several authors have dealt with the rheological problems encountered in the flow of a non-Newtonian fluid along an annulus, which a drilling situation effectively represents [105-110], and used various models combining the properties of a Bingham fluid and power law fluid to do so, with varying degrees of success. The most successful model would appear to be that quoted earlier for cement pastes [97], namely

$$\tau = A(q + c)^B$$

A three constant Oldroyd model has also been used [111,112] to investigate discharge rates and torque on the inner pipe produced by helical flow. A numerical method was used, general enough to be used with other models, but the predictions of the theory were not compared with experimental results.

Despite the fact that theoretical work has utilised the Bingham model to describe the behaviour of a drilling mud, recent unpublished work [113] with a rheogoniometer was unable to find positive evidence of a yield stress in two commercial muds tested. The same muds gave no evidence of normal stresses (within the limits of experimental error) when tested on a torsional balance rheometer.

In industrial practice, the rheology of a drilling mud is monitored with a Fann viscometer (Figure 35),

Fig. 35   Fann viscometer

a multipoint rotational instrument. More specific details of various models of this viscometer are given in a Bulletin issued by the American Institute of Petroleum [114], together with instructions for processing the data produced by it. Factors so calculated include:

   (i) Average bulk velocity in a pipe or annulus
  (ii) Shear rate at the pipe or annulus wall
 (iii) Effective viscosity in a pipe or annulus
  (iv) Reynolds number in a pipe or annulus
   (v) Friction factor in a pipe or annulus
  (vi) Friction loss pressure gradient in a pipe
         or annulus
 (vii) Hydrostatic pressure gradient
(viii) Circulating pressure gradient
  (ix) Equivalent circulating density.

3.6 Pharmaceuticals

Pharmaceutical materials are prepared in a variety of forms; a significant proportion will be produced as pastes or ointments, which we might class as semi-solids, or as tablets and powders, and such materials we may consider to lie outside the scope of this chapter. Three categories of material do, however, deserve our attention:

Particulate suspensions, *per se*, such as suspensions of aluminium hydrate intended to be taken internally, or lotions for external use, such as calamine;
Particulate suspensions intended for administration by injection, such as suspensions of antibiotics;
Liquid-in-liquid suspensions (emulsions) which could either carry a medicament to be taken internally, or some compound for external application, as would be the base in a cream or emulsion-based lotion.

It is unlikely that the rheology of a material intended for oral consumption will be a significant factor in determining whether or not it can be swallowed; rheological properties will more probably be linked to the instruction "Shake the bottle". Experience determines the level and frequency at which a medicament should be taken, and unless the concentration of medicament in suspension form is uniform and remains so as the bottle is progressively emptied, the dose rate is not necessarily uniform and an overdose may unwittingly be taken. Such a state of affairs could reasonably occur in a well dispersed suspension whose particles settled out rapidly and gave a hard packed sediment on the bottom of the bottle, which would not readily disperse. This case of redispersion is important - elderly patients might find it difficult to give a bottle a vigorous and prolonged shaking.

The structure required to maintain the particles in suspension is achieved by addition of a soluble polymer, such as a natural gum, derivatives of cellulose (such as CMC) or derivatives of alginic acid (e.g. sodium alginate), or it might consist of an inorganic pigment of high specific surface area, capable of imparting thixotropic structure to the suspension in much the same way as bentonite does to a drilling mud. One commonly used inorganic material is Veegum, a complex magnesium aluminium silicate containing a small proportion of metal oxides. Materials used to impart structure are known as suspending agents.

The pseudoplastic nature of CMC suspensions has already been illustrated (Figure 34), while similar behaviour has been demonstrated in the case of alginate based materials, and the power law shown to apply [115].

Other work, in which solutions of a variety of polymeric suspending agents were measured with a capillary viscometer and falling sphere viscometer [116] established that the relative viscosity of such a solution, $\eta_{rel.}$, could be related to the concentration of suspending agent by a relation

$$\log \left(\frac{\eta_{rel.}}{c} \right) = K$$

where K is a constant whose value depends on the molecular weight of the polymer employed.

Flow curves of freshly prepared Veegum suspensions

exhibit pseudoplastic thixotropic behaviour, but a yield stress develops as the suspension ages. The build up of viscosity with time can be represented by the equation [117]

$$\log \eta = a \log t + b$$

where η is the apparent viscosity, t is time, and a and b are constants whose values depend on ageing time, temperature and the shear rate of measurement.

3.6.1 Suspensions for injection

Here, the rheological properties of the material concerned can be of significance in determining the ease or otherwise of administration. A simple example will illustrate this. It is easily calculated that if 2 ml of fluid are passed through a capillary of 0.2 mm bore in 10 seconds, then the shear rate at the wall is 2.5×10^5 s^{-1}. If the bore is increased to 0.5 mm then the shear rate drops to 1.6×10^4 s^{-1}. What the actual bore of a hypodermic needle is depends on a number of factors, but one can say that the shear rates developed in it are significant, and that the rate of change of shear rate between syringe barrel and needle will be quite high. Consequently one might anticipate difficulty in attempting to inject a suspension which was markedly shear thickening or showed significant elastico-viscous behaviour, or possessed too high a yield stress.

Work done on aqueous suspensions of procaine penicillin G with a Hercules viscometer [118] gave curves showing both a significant yield stress and marked thixotropic hysteresis. Rate of structural recovery was rapid since, after a few minutes standing, the up curve of "rested" material coincided with that of the original. This combination of yield stress, rapid breakdown and rapid recovery is useful, in that it enables a fairly concentrated suspension to be injected readily and yet (because of the structural recovery) gives a compact deposit of the antibiotic in the blood vessel, which then releases antibiotic slowly into the blood stream and over a prolonged period. Storage of such suspensions at temperatures below

body temperature caused an increase in viscosity and
yield value [119], the increase being more pronounced
with increased storage temperature. Storage for three
weeks at 37°C produced a material too viscous to be
injected.

3.6.2 Emulsion-based suspensions

The subject of emulsions is a considerable one, and
for wider reading the reader can justifiably be referred to one of the specialised texts [120,121,122].

An emulsion is essentially a stable suspension of
droplets of one liquid in a second liquid with which
it is immiscible; since the two liquids concerned are
commonly water and an oily liquid, emulsions containing these constituents are referred to as "oil in
water" (O/W) or "water in oil" (W/O), depending on
which constituent is dispersed in which. A stable
emulsion has the advantage of presenting its constituents uniformly distributed over the whole sample (as
a stable suspension would), and has the further advantage that it enables some control to be exerted on the
rate of medicaments into the system if administered
internally or to the skin if applied as a cream.

Emulsions are often shear thinning fluids. Viscosity
depends, reasonably enough, on the concentration of
the dispersed phase present and also on the particle
size of that phase. It has been shown [123] that for
W/O emulsions viscosity increases significantly as the
particle size of the dispersed phase decreases, but
was less pronounced for O/W emulsions, and was only
apparent there for fractional volume concentrations
above 0.5. The results point to one further factor
controlling the rheological properties of an emulsion,
namely, the rheological properties of the interface
between the continuous and the dispersed phase.

Both classes of emulsion can show thixotropic behaviour and exhibit a yield stress; work reported on W/O
emulsions of white petrolatum in water, stabilised
with a sorbitan monooleate surfactant [124] and of O/W
emulsions stabilised with potassium arabate [125] could
be cited as examples.

In the case of an O/W emulsion, a fatty alcohol is often incorporated in the oil phase, and the concentration of that alcohol can have a significant influence on the rheology of the emulsion [126]. There is evidence that when the fatty alcohol exceeds its saturation concentration in the oil phase, the excess can move into the water phase, complex with the surfactant present there to form a liquid crystal, which then appears in the system as a gel. Work done with rotational viscometers and a rheogoniometer [127,128] using a similar liquid crystal phase formed by the reaction of cetyl alcohol with an aqueous solution of sodium dodecyl sulphate, found no evidence of a yield stress, but did demonstrate the presence of a hysteresis loop. Recovery of the structure indicated by that loop, however, was never obtained, so true thixotropy was not present. Loss tangent and creep compliance measurements indicated the presence of elastico-viscous behaviour.

The comments in this section are intended to serve only as a sketch of the type of rheological behaviour which an emulsion might exhibit. The complexity of that behaviour - in common, one might say, with that of most suspensions - is clear.

3.6.3 Rheological control of pharmaceuticals in industry

Preparation of pharmaceutical suspensions seems to have something in common with that of paper coating mixes, in that surfactants and soluble polymers will often be present in addition to the suspended phase and suspending liquid. Their non-Newtonian properties will be equally important as far as makedown, pumping and screening are concerned, operations which are as relevant to the preparation of pharmaceuticals as to paper coating. For control, rotational instruments such as the Brookfield, Epprecht Rheomat or Haake Rotovisco might be used; the philosophy of using a one-point measurement to control a non-Newtonian suspension of given formulation is known to the pharmaceutical industry also.

4. CONCLUSIONS

Research has amply demonstrated the complex nature
of the rheology of industrial suspensions; suspensions
can be used to demonstrate virtually every model type
of rheological behaviour known. The rheology is influenced by the shape, size and concentration of the
suspended particles, by the surface chemistry of the
solid, by its interaction with the liquid phase and -
in the case of an emulsion - by the rheology of the
interface between the components.

The rheological behaviour is also determined by the
shear rate of measurement - a paper coating mix, for
instance, may show shear-thinning behaviour at one
shear rate and shear thickening at another, higher,
shear rate.

Industrial control of such materials, however, is
almost universally simple, frequently elementary and
occasionally non-existent. There are several reasons
for this; we could cite the following:

(a) It is often difficult to translate identifiable
complex rheology into terms suitable for control or
application purposes.
(b) Simple methods (such as one point measurements
with a Brookfield) are often found by experience to be
adequate for control purposes. Why, then, invest in
an expensive instrument, requiring skill to operate it
and interpret its results?
(c) Suspensions, by their nature, and the conditions
in which they are used, tend to be "dirty" - complex
rheometers are not necessarily suited to such conditions. A rheogoniometer may be useful for establishing
the rheological properties of a cement paste, for instance, but a building site would hardly provide a
suitable location in which to use it.

No doubt there are other reasons, but the level of
rheological experience provided by the research laboratory on suspensions and that demanded by and utilised
by the shop floor is such that one might say that
"between us and you there is a great gulf fixed: so

that they which would pass from hence to you cannot;
neither can they pass to us, that would come from
thence" [129]. Progressive bridging of this gulf
would surely be beneficial to both parties.

ACKNOWLEDGEMENTS

I acknowledge with thanks the consent of the Board
of English China Clays International to contribute
this chapter.

I would also wish to express my thanks to the following for the helpful discussions which I have had with
them: Professor B.W. Barry, University of Bradford;
Dr. T.E.R. Jones, Plymouth Polytechnic; Dr. G.
Tattersall, University of Sheffield; Mr. F. Moore,
B.C.R.A., Stoke on Trent; Dr. B. Warburton, School
of Pharmacy, University of London.

My thanks are also due to Mr. F. Moore for permission
to reproduce Figure 25 and to Wykeham Farrance
Engineering Limited, Slough, for supplying the photograph for Figure 32.

REFERENCES : CHAPTER 7

[1] A. Einstein, A. Phys. Leipzig, 19 (1906) 289

[2] A. Einstein, ibid, 34 (1911) 591

[3] G.B. Jeffery, Proc. Roy. Soc., 102 (1923) 161

[4] J.M. Burghers, Second Report on Viscosity and
     Plasticity, N.V. Noord-Hollandsche Uitgevers-
     maatschappij, Amsterdam, 1938, p. 113

[5] V. Vand, J. Phys. Coll. Chem., 52 (1948) 277

[6] F. Eirich, M. Bunzyl and H. Margaretha,
     Kolloid Z., 74 (1936) 276

[7] E. Guth and R. Simha, Kolloid Z., 74 (1936) 266

[8] R. Simha, Kolloid Z., 76 (1936) 16

[9] M. Kunitz, J. Ges. Physiol., 9 (1925) 715

[10] H. De Bruijn, Rec. Trav. Chim. Pays-Bas, 61
     (1942) 863

[11] J.V. Robinson, J. Phys. Coll. Chem., 53 (1949)
     1042

[12] M.J. Mooney, J. Coll. Interface Sci., 6 (1951) 182

[13] J.C. Brodnyan, Trans. Soc. Rheol., 3 (1959) 61

[14] W. Windle and K.M. Beazley, TAPPI, 51(8) (1968)
     340

[15] K.M. Beazley, Ph.D. Thesis, University of
     London, 1966, p. 95

[16] S.A. Bullivant, Ph.D. Thesis, Plymouth Poly., 1977

[17] G.A. Gill, TAPPI Monograph No. 38, "Paper Coating
     Pigments", Technical Association of the Pulp
     and Paper Industry, Atlanta, 1976, p. 178-90

[18]   L. Pauling, Proc. Nat. Acad. Sci., 16 (1930) 123

[19]   L. Pauling, Proc. Nat. Acad. Sci., 16 (1930) 578

[20]   P.A. Thiessen, Z. Electrochemie, 48 (1947) 675

[21]   R.K. Schofield and H.R. Samson, Disc. Faraday
       Soc., 18 (1954) 135

[22]   R.K. Schofield and H.R. Samson, Clay Min. Bull.,
       2 (1953) 45

[23]   U. Hofmann, F.W. Schembra, M. Schatz,
       D. Scheurlen and H. Friedrich, Kolloid Z.,
       216-7 (1967) 370

[24]   A.S. Michaels and J.C. Bolger, Ind. Eng. Chem.
       Fundamentals, 1(3) (1962) 153

[25]   E.J.W. Verwey and J.Th. Overbeek, "Theory of the
       Stability of Lyophobic Colloids", Elsevier,
       New York, Amsterdam, London, Brussels, 1948

[26]   B. Derjaguin and L.D. Landau, Acta physicochim.
       U.R.S.S. 14 (1941) 635

[27]   B. Derjaguin and L.D. Landau, J. Exp. Theor.
       Phys., USSR, 11 (1941) 802

[28]   S.R. Dennison and G.L. Toms, TAPPI, 50(10) (1941)
       502

[29]   S.R. Dennison, Woch. Papierfabr. 102 (1974) 156

[30]   F.L. Norton and A.J. Johnson, J. Am. Ceram. Soc.,
       27 (1944) 77

[31]   H.H. Macey, Trans. Brit. Ceram. Soc., 41 (1941)
       73

[32]   A. De Waele and E.W.J. Mardles, Proc. 1st Int.
       Cong. Rheol. Holland, 1949. II. 166 - III. 172

[33]   K.M. Beazley, Trans. Brit. Ceram. Soc., 63(9)
       1964, 451

408

[34] A.W. Helz, J. Am. Ceram. Soc., 22 (1939) 289

[35] B.K. Adsell, Tech. Asssoc. Papers, 31 (1948) 193

[36] F.H. Norton, A.L. Johnson and W.G. Lawrence, J. Am. Ceram. Soc., 27 (1944) 149

[37] N. Millman, TAPPI, 47 (1964) 168A

[38] H.H. Morris, P. Sennett and R.J. Drexel, TAPPI, 48 (1965) 92A

[39] A. Gabrysh, H. Eyring and I. Cutler, J. Am. Ceram. Soc. 45 (1962) 334

[40] Van Olphen, "An Introduction to Clay Colloid Chemistry", Interscience, New York, 1963, Chap. 3

[41] R.E. Brociner, TAPPI, 48(9) (1965) 110A

[42] R.E. Brociner, Woch. Papierfabr., 23/24 (1971) 1003

[43] K. Kraft, Woch. Papierfabr., 23/24 (1971) 1008

[44] R.W. Williams, Rheol. Acta 18 (1979) 345

[45] L. Frylof, Svensk, Farm. Tids., 30 (1961) 753

[46] N.O. Clark, Trans. Brit. Ceram. Soc., 49 (1950) 409

[47] N.O. Clark and R.J. Booth, Proc. Tech. Sec. B.P.B.M.A., 32 (1951) 89

[48] K.M. Beazley, J. Coll. Int. Sci., 41(1) (1972) 105

[49] K.M. Beazley, TAPPI, 50(3) (1967) 151

[50] Dr. Nötzel, Woch. Papierfabr., 20 (1978) 779

[51] R. Bohmer, Svensk, Papperstid., 9(15) (1964) 347

[52] S.N. Hussain, W.R. Farrall, M.M. Dalfin and
J.P. Gunning, TAPPI, 53(1) (1970) 96

[53] C.R. Price and R.W. Hagemayer, TAPPI, 61(5)
(1978) 47

[54] S.R. Dennison, TAPPI, 62(1)(1979) 65

[55] L. Huggenberger, W. Kogler and M. Arnold,
TAPPI, 62(5) (1979) 37

[56] D. Borruso, A. Croce and A. Seves, Ind. della
Carta, 7(3) (1969) 100

[57] F. Moore and L.J. Davies, Trans. Brit. Ceram.
Soc., 55 (1956) 313

[58] F. Moore, Trans. Brit. Ceram. Soc. 58 (1959) 470

[59] F. Moore, Rheology of Ceramic Systems, Institute
of Ceramics, Stoke on Trent, 1965, p. 35

[60] Ceramic Tests, E.C.C. International, St. Austell,
Cornwall.

[61] A.A. Robertson and S.G. Mason, TAPPI, 40(5)
(1957) 326

[62] N.O. Clark, W. Windle and K.M. Beazley, TAPPI,
52(11) (1969) 2191

[63] K.M. Beazley and N. Climpson, TAPPI, 53(12)
(1970) 2227

[64] W. Windle, K.M. Beazley and N. Climpson, TAPPI,
53(12) (1970) 2232

[65] W. Windle and K.M. Beazley, Woch. Papierfabr.
No. 10, of 1973, 332

[66] E. Böhmer and J. Lute, Svensk. Papperstids.,
69(17) (1966) 610

[67] E. Böhmer and J. Lute, Svensk. Papperstids.,
68(20) (1965) 711

410

[68] H.P. Gore, TAPPI, 49(11) (1966) 473

[69] S.F. Kurath and D.G. Berge, TAPPI, 53(4) (1970) 646

[70] S.F. Kurath, TAPPI, 52(1) (1969) 92

[71] S.F. Kurath and W.S. Larson, TAPPI, 54(9) (1971) 1509

[72] W. Windle and K.M. Beazley, TAPPI, 50(1) (1967) 1

[73] W.J. Follette and E.W. Fowells, TAPPI, 43 (1960) 953

[74] W.C. Bliesner, TAPPI, 54(10) (1971) 1673

[75] L.L. Turai, TAPPI, 54(8) (1971) 1315

[76] E. Böhmer, Norsk Skogind, Aug. 1968, 258

[77] E. Böhmer, Norsk Skogind, Nov. 1969, 308

[78] H. Pummer, Allg. Pap. Rundsch. 1968, 1064, 1085, 1138, 1487-88, 1553-55

[79] R.P. Mottiot, Ind. della Carta, 9(4) (1971) 194

[80] F. Mancosu, Cellul. Carta, 27(12) (1976) 19

[81] D. Eklund and S.J. Kahila, Woch. Papierfabr. 17 (1978) 661

[82] D. Eklund, S.J. Kahila and D. Obetko, Woch. Papierfabr. 18 (1978) 709

[83] J.W. Smith, R.T. Trelfa and H.O. Ware, TAPPI, 35 (1950) 212

[84] N.F. Astbury, Trans. Brit. Ceram. Soc., 62 (1963) 1

[85] E.C. Bloor, Trans. Brit. Ceram. Soc., 56 (1957) 423

[86] J.W. Whittemore, J. Am. Ceram. Soc., 18 (1935) 352

[87] I.A. Alperovitch, P.I. Berenstein and G.S. Bloch. Stek. Keram, 10 (1953) 22

[88] M. Juul Hvorslar, Ingeniorvidenskabelige Skrifter, A. No. 45 (1937) Danmarks Natur-videnskabelige Samfund, Copenhagen.

[89] Le Chatelier, Experimental Researches on the Constitution of Hydraulic Mortars. Translated by J.L. Mack, New York, 1905

[90] Le Chatelier, Compt. Rend., 96 (1883) 1056

[91] A.E. Törnebohm, Ton. Ind. Zeit. (1897) p.1148

[92] S. Taylor, M. Phil. thesis, Plymouth Polytechnic, 1979

[93] G.H. Tattersall, Brit. J. Appl. Phys., 6(5) (1955) 165

[94] C.R. Dimond, Ph.D. thesis, University of Sheffield, 1975

[95] C.R. Dimond and G.H. Tattersall, Proceedings of a Conference, 8-9 April 1976, Sheffield. Cement & Concrete Ass., London, 1976, p.118-33

[96] T.E.R. Jones and S. Taylor, Magazine of Concrete Research, 29(101) (1977) 207

[97] R.E. Robertson and H.A. Stiff, Jr., J. Soc. Petr. Eng., 16(1) (1976) 31

[98] T.E.R. Jones and S. Taylor, Sil. Ind., 4/5(1978) p. 83

[99] T.E.R. Jones, Plymouth Polytechnic, private communication

[100] K. Endell, U. Hofmann and B. Wilm, Ber. Deut. Ker. Ges. 14, 1935, 407

412

[101] C.E. Marshall, Z. Krist., 91 (1935) 433

[102] E. Magdefrau and U. Hofmann, Z. Krist., 98
       (1937) 299

[103] S.B. Hendricks, J. Geol., 50

[104] D. Eklund, Paperi ja Puu, No. 9 (1976) 558

[105] A.G. Fredrickson and R.B. Bird, Ind. Eng.
       Chem., 50(3) (1958)

[106] D.W. Dodge and A.B. Metzner, A.I.Ch.E.Proc.,
       5(2) (1959) 189

[107] B. Le-Fur and M. Martin, J. Fl. Mech., 30(3)
       (1967) 449

[108] M.R. Annis, J. Pet. Tech. (1967) 1074

[109] R.E. Walker and D.E. Korry, J. Pet. Tech.
       (1974) 167

[110] R.W. Flummerfelt, S.P.E.J. (1975) 169

[111] J.G. Oldroyd, Proc. Roy. Soc., A245 (1958) 278

[112] J.G. Savins and G.C. Wallich, A.I.Ch.E.Proc.,
       12(2) (1966) 357

[113] M.A. Lockyer and T.E.R. Jones, Plymouth
       Polytechnic, private communication

[114] API Bulletin on 'The Rheology of Oil Well
       Drilling Fluids' issued in draft form,
       Nov. 1977

[115] S.P. Kabre, H.G. De Kay and G.S. Banker,
       J. Pharm. Sci., 53 (1964) 492

[116] K. Munzl and K. Schaub, Pharm. Acta Helv., 36
       (1961) 647

[117] J.H. Wood, Amer. Perfumer., 76 (1961) 37

[118] S.S. Ober, H.C. Vincent, D.E. Simon and
      K.J. Frederick, J. Am. Pharm. Assoc., 47
      (1958) 667

[119] J.C. Boylan and R.L. Robinson, J. Pharm. Sci.,
      57 (1968) 1796

[120] P. Sherman, Emulsion Science, Academic Press,
      1968

[121] P. Sherman, Industrial Rheology, Academic
      Press, 1970

[122] P. Sherman (Ed.), Rheology of Emulsions,
      Pergamon Press, London, 1963

[123] P. Sherman, Proc. 3rd Intern. Congr. Surface
      Activity II,

[124] H.B. Kostenbauder and A.N. Martin, J. Am.
      Pharm. Assoc. 43 (1954) 401

[125] E. Shotton and S.S. Davis, J. Pharm. Pharmacol.
      20 (1968) 780

[126] F.A.J. Talman, P.J. Davies and E.M. Rowan,
      J. Pharm. Pharmacol. 19 (1967) 417

[127] B.W. Barry and E. Shotton, J. Pharm. Pharmacol.
      19 suppl. (1967) 1105

[128] B.W. Barry and E. Shotton, J. Pharm. Pharmacol.
      19 suppl. (1967) 1215

[129] St. Luke, 16. 26

INDEX

Anti-thixotropy, 3,84
Apparent viscosity, 7,8,11,37,53,72,73,74,80,83,94,
     124,126,128,131,165,224,236,237,260,262,264,287,
     288,345,376,397,401

Band viscometer, 327,370
Bingham plastic, 8,11,378,397
Boger liquid, 13,16,18
Bridgman falling-ball technique, 153
Brookfield viscometer, 12,14,41,42,51,65,67,132,146,
     149,164,367,368,370,378,383,384,386,403,404

Cannon Fenske viscometer, 146
Cannon mini-rotary viscometer, 149
Capillary viscometer, 54,62,83,124,128,132,137,140,
     145,146,149,150,153,165,166,169,170,240,241,242,
     312,318,370,373,382,400
Cold-cranking simulator, 125,152
Combined steady and oscillatory shear, 19,20,78,186
Complex viscosity, 17
Concentric cylinder viscometer, 49,50,51,54,64,65,140,
     149,150,152,157,233,253,308-310,318,320,321,373,
     389
Cone and plate flow, 11,12,51,76,169,171,235,308,310-
     313,320,324,327,390
Contraves viscometer, 51,84,322,386
Converging flow, 23,53,98,136,247,267

415